本书的出版得到以下林业公益性行业科研专项的支持：
中国森林对气候变化的响应与林业适应对策研究
湘鄂石漠化地区常绿阔叶林恢复研究

Tropical Nature Plants in
Nanling Mountains Region of Hunan

湖南南岭地区热带性植物

徐永福　曹福祥　喻勋林 ▪ 著

中国林业出版社

著　者：徐永福　曹福祥　喻勋林

摄　影：喻勋林　徐永福　曹基武　曹福祥　付　英　张贵志　黄　娟　罗开文　邹　滨

顾　问：祁承经

Writers: XU Yong-fu, Cao Fu-xiang, YU Xun-lin

Photographers: YU Xun-lin, XU Yong-fu, CAO Ji-wu, CAO Fu-xiang, FU Ying, ZHANG Gui-zhi, HUANG Juan,
LUO Kai-wen, ZOU Bin

Consultant: QI Cheng-jing

图书在版编目（CIP）数据

湖南南岭地区热带性植物：汉英对照 / 徐永福, 曹
福祥, 喻勋林著. -- 北京：中国林业出版社, 2012.10
ISBN 978-7-5038-6765-1

Ⅰ. ①湖… Ⅱ. ①徐… ②曹… ③喻… Ⅲ. ①南岭 –
热带植物 – 湖南省 – 图集 Ⅳ. ①Q948.526.4-64

中国版本图书馆CIP数据核字(2012)第228883号

出　　版：中国林业出版社
　　　　　（100009　北京西城区德内大街刘海胡同 7 号）
电　　话：(010) 8322 9512
发　　行：新华书店
印　　刷：北京顺诚彩色印刷有限公司
版　　次：2012 年 10 月第 1 版
印　　次：2012 年 10 月第 1 次
成品尺寸：185mm × 260mm
印　　数：1000 册
印　　张：15.25
字　　数：300 千字
定　　价：150.00 元

前　言

　　南岭是我国南部的一道天然屏障，狭义的南岭主要由越城岭、都庞岭、萌渚岭、骑田岭和大庾岭所组成。山间多山隘和盆地，地形较破碎而复杂。分布范围很广，横跨福建西南部、江西南部、湖南南部、广东北部和广西东北部。在湖南境内的南岭地区地处南岭中段，位于北纬24º39′～26º55′，东经109º58′～114º14′，除大庾岭外，其余四岭均在湖南境内或边界上，其中包括零陵地区的永州市、冷水滩市、祁阳县、东安县、双牌县、道县、江永县、江华县、蓝山县、宁远县、新田县，郴州地区的郴州市、汝城县、宜章县、郴县、桂阳县、临武县、嘉禾县，邵阳地区的新宁县和城步县，共计有20个县（市），土地总面积为39 462km²，约占湖南省国土面积的18.63%。

　　本区属中亚热带季风湿润气候区，是南北气团交汇地带，南岭山地对北来的寒潮、南来的季风都有一定的阻隔作用。年平均气温可达17.6～18.6℃，年日照时数为1 600～1 750h，无霜期281～311d，≥10℃的积温为5 530～5 800℃，年降水量为1 200～2 000mm，为湖南省水热最丰富的地区，孕育着丰富的植物资源，其中包含着许多热带性植物。本书将热带性植物定义为主产南亚热带或热带地区而延伸至湖南南岭地区且主要垂直分布于海拔800m以下的野生植物。

　　在林业公益性行业科研专项"中国森林对气候变化的响应与林业适应对策研究"的大力支持下，编者通过对湖南南岭地区，特别对该区特殊地形下形成能避风的"暖窝子"进行热带性植物调查发现，湖南南岭地区分布有许多严格的热带植物区系成分，如买麻藤科、莲叶桐科、粘木科、古柯科、使君子科、山榄科、茶茱萸科、五列木科、野牡丹科(灌木种)、梧桐科、罗汉松科(野生种)、蒟蒻薯科、芭蕉科、棕榈科等。其中所包含的属和种局限分布于中亚热带南部亚地带，绝少见于北部亚地带，多以南岭北缘为其分布区的北界，这些属和种可作为中亚热带南、北亚地带区分的标志和依据。

本书依据长期野外调查结果，在约500种热带性植物中，现挑选出主要的、典型热带性植物202种，用中英文对其形态特征、生存环境、分布产地及主要用途进行论述，并配有相应的彩色照片，图文并茂。特别对首次发现于湖南的南亚热带植物和典型热带性植物的天然分布最北缘给予报道，这将对进一步研究南岭地区及湖南植物区系性质奠定重要基础，同时对研究本区热带性植物迁移及经济热带植物引种栽培提供依据。此外，在全球气候变化（变暖）的背景下，关注热带性植物分布北缘的分布动态，将对监测气候变化提供重要佐证。本书植物编排蕨类植物的科按秦仁昌系统排列，裸子植物的科按郑万均系统排列，被子植物的科按哈钦松系统排列。

　　在进行野外调查时得到汝城九龙江国家森林公园、莽山国家自然保护区、蓝山国家森林公园、九嶷山国家森林公园、舜皇山国家森林公园、城步两江峡谷国家森林公园、都庞岭国家级自然保护区、黄桑国家自然保护区、通道县林业局等单位和部门的鼎力支持，特此致谢！

　　由于热带性植物调查是一个长期的过程，加之编者水平有限，书中将会存在一定的缺点和错误，恳请读者批评指正。

编者

2012年6月

▋ Foreword

 Nanling Mountain is a natural barrier in the south of China on physical geography. On narrow sense, Nanling is mainly consisted of Yuechengling, Doupangling, Mengzhuling, Qitianling and Dayuling. Mountain passes among ravine multi-cols and river basins, and the terrain is more fragmented and complex. The distribution is wide as which stretches cross five provinces including southwestern Fujian, southern Jiangxi, southern Hunan, northern Guangdong and northeastern Guangxi. It is the middle part of Nanling in Hunan's territory which located at latitude 24°39'−26°55 ', longitude 109°58'−114°14'. Except for Dayuling, other 4 ridges are located in or on the boundary of Hunan, that covered 20 counties which are Yongzhou city, Lengshuitan city, Qiyang county, Dongan county, Shuangpai county, Daoxian, Jiangyong county, Jianghua county, Lanshan county, Ningyuan county, Xintian county in Lingling area, and Chenzhou city, Rucheng county, Yizhang county, Chenxian, Guiyang county, Linwu county, Jiahe county in Chenzhou area, and Xinning county and Chengbu county in Shaoyang area. The total area is 39 462km^2, which is about 18.63% of the total Hunan area.

 This area belongs to the humid subtropical monsoon climate zone, and it is the north and south air mass connection region. Nanling Mountains can block the cold wave from north and the monsoon from south to some extent. By which the annual mean temperature may reach 17.6−18.6℃, the annual sunshine hours to 1 600−1 750 hours, the frost-free period 281−311 days, the accumulated temperature ≥ 10℃ for 5 530−5 800℃, precipitation in the 1 200−2 000mm. Owing to the most abundant hydrothermal condition in southern Hunan, it is rich in typically tropical floristic elements and the numerous tropical plants. The definition of tropical nature plants in this book is the wild plants which are mainly growing in southern subtropical or tropical areas and extended to the Hunan Nanling region where mainly below alt. 800 m.

 In the strong support of Special Research Program for Public-welfare Forestry

'Responses of forests to climate change and adaptive strategy of forestry in China', the editors carried out the tropical nature plants investigation through Hunan Nanling area, especially to this area where under the extreme terrain forms 'local warm habitat' which can take shelter from wind, and the results showed that there are many strictly tropical plants floristic distribution in Hunan Nanling area, such as Gnetaceae, Hernandiaceae, Ixonanthaceae, Erythroxylaceae, Combretaceae, Sapotaceae, Icacinaceae, Pentaphylacaceae, Melastomataceae(shrub species), Sterculiaceae, Podocarpaceae(wild species), Taccaceae, Musaceae, Palmae, and so on. The plants belong to the genera and species in these families are limitedly distributed in the middle subtropics southern sub-zone area, and rarely seen in the middle subtropics northern sub-zone ones, and usually taking the north edge of Nangling as its northern boundary of distribution area. These genera and species can be seen as a characteristics and the basis to distinct north sub-zone and south one in middle subtropics.

Based on long-term field survey results in this book, we select the main typical 202 species among about 500 species of tropical nature plants, and describe their morphological characteristics, survival environment, distributed habitat and main uses of tropical nature plants in Hunan Nanling area with bilingual English-Chinese languages and excellent pictures and essay, especially reporting the south subtropical plants which were first discovered in Hunan, and describing the northern nature distribution boundary of typical tropical nature plants. The results will establish an important foundation for further study of the flora nature in Nanling and Hunan. Meanwhile, the results also provide the basis for studying the migration of tropical nature plants and introduction cultivation of the economic tropical nature plants. Additionally, in changed (warming) global climate background, paying attention to dynamic distribution on northern margin of tropical plant distribution will provide important evidence for monitoring climate change. Plants arrangement in this book: Ferns according to Qin Renchang system, Gymnosperms according to Zheng Wangjun system, Angiosperms according to Hutchinson system.

During this field survey period, we have got the strong support from many units and departments such as Rucheng Jiulongjiang National Forest Park, Mangshan National Nature Reserve, Lanshan National Forest Park, Jiuyishan National Forest Park, Shunhuangshan National Forest Park, Chengbu Liangjiang Valley National Forest Park, Dupangling Nature Reserve, Huangsang National Nature Reserve, Tongdao County Forestry Bureau and so on, and hereby express our gratitude.

As to the long-term survey and the editors' limited level, there is some shortcomings and mistakes, and the correction is welcomed sincerely.

Editors
June, 2012

目 录 Contents

本书的出版得到以下林业公益性行业科研专项的支持：
中国森林对气候变化的响应与林业适应对策研究
湘鄂石漠化地区常绿阔叶林恢复研究

湖南南岭地区热带性植物

Tropical Nature Plants in
Nanling Mountains Region of Hunan

福建观音座莲 *Angiopteris fokiensis* Hieron.

多年生草本，高1.5~3m。根状茎块状，直立。叶簇生；叶柄粗壮，多汁肉质，长50~70cm；基部有肉质托叶状附属物；叶片阔卵形，2回羽状；羽片互生，狭长圆形，基部略狭，宽14~18cm；小羽片平展，披针形，先端渐尖，基部近截形或近圆形，边缘有浅三角形锯齿，叶脉单一或二叉。孢子囊群长圆形，长约1mm，着生于距叶缘0.5~1mm处，通常由8~12个孢子囊组成。

产通道、江华、东安、宜章。生林下沟谷或溪边。分布于重庆、福建、广东、广西、贵州、海南、香港、湖北、江西、四川、云南。日本南部亦产。株形优美，系优良的大型观赏蕨类，适宜于大盆栽培。

Perennial herbs, 1.5-3m tall. Rhizomes blocky, erect. Leaves tufted; petioles stout, jsucculent, 50-70cm long; base with succulent appendages; leaves broadly ovate, bipinnate; pinnae alternate, narrowly oblong, base slightly narrow, 14-18cm wide; pinnules spreading, lanceolate, apex acuminate, base subtruncate or suborbicular, margin shallowly triangular serrated. Vein single or binary. Sori oblong, ca.1mm long, borne in 0.5-1mm from margin, usually composed of 8-12 sporangia.

Distributed in Hunan(Tongdao, Jianghua, Dongan, Yizhang), Chongqing, Fujian, Guangdong, Guangxi, Guizhou, Hainan, Hongkong, Hubei, Jiangxi, Sichuan and Yunnan, also in South Japan. Grows in valleys under the forest or along brooks. As a kind of fine large ornamental ferns, the individual plant is beautiful and suitable for large bowl cultivation.

华南紫萁 *Osmunda vachellii* Hook.

多年生草本，高达1m。根状茎圆柱形，高出地面，顶部叶簇生。羽片2型，叶片矩圆形，厚纸质，光滑，长40～90cm，宽20～30cm，1回羽状，中部以上的羽片不育，披针形或条状披针形，基部狭楔形，长渐尖，边缘全缘；侧脉1～2回分叉；下部羽片常能育，窄条形，宽4mm，深羽裂，有宽缺刻，裂片两面沿叶脉密生孢子囊，形成圆形小穗，排列在羽轴两侧。

产江华、东安、汝城、武冈、宜章。生低海拔灌丛及林下。分布于四川、贵州、云南、浙江、福建、广东、广西、香港、海南。印度、缅甸、越南亦产。

Perennial herbs, up to 1m tall. Rhizomes cylindrical, above the ground, leaves tufted at the top of rhizomes. Pinnae dimorphic, leaf blade oblong, thick papery, smooth, 40-90 × 20-30cm, pinnate, pinna sterile above middle, lanceolate or strip-lanceolate, base narrowly cuneate, long acuminate, margin entire; lateral veins 1-2 bifurcated; lower pinnae often fertile, narrowly bar, 4mm wide, deeply pinnatifid, with wide toothed, sporangia densely borne on veins at both surfaces of lobes, forming rounded spikelets, arranged in plume axis on both sides.

Distributed in Hunan(Jianghua, Dongan, Rucheng, Wugang, Yizhang), Sichuan, Guizhou, Yunnan, Zhejiang, Fujian, Guangdong, Guangxi, Hongkong and Hainan, also in India, Burma, Vietnam. Grows in shrubs and in forests at low altitude.

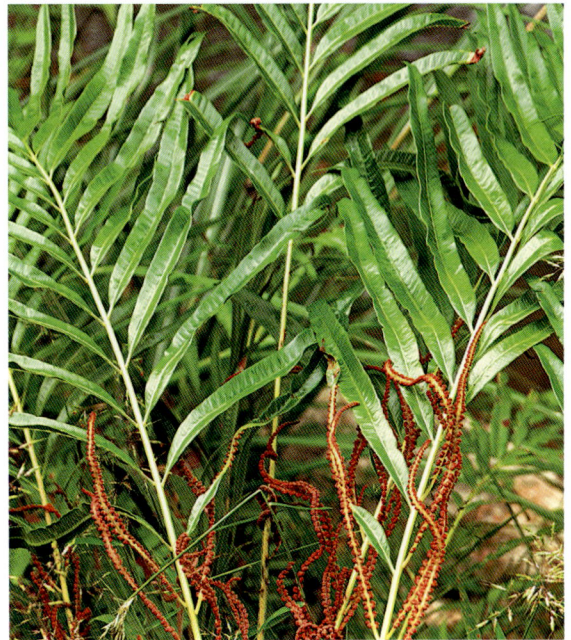

小叶海金沙 *Lygodium microphyllum* (Linn.) Sw.

蔓生草本。茎纤细，2回羽状；羽片多数，相距7～9cm。叶2型，无毛；不育叶长圆形，长7～8cm，宽4～7cm，单数羽状，羽片小，心形或卵状三角形，长1.5～3cm，钝头，基部心形，边缘有短锯齿。能育叶同形，长8～10cm，宽4～6cm，小羽片三角形或卵状三角形。孢子囊穗条形，长3～6mm，无毛，每穗有孢子囊5～8对，排列于叶缘。

产通道、江华、宜章。生低海拔山坡及山谷灌丛中。分布于福建、广东、广西、海南、香港、江西、台湾、云南南部。亚洲热带其他地区亦产。

Trailing herbs. Stem slender, bipinnate; pinnae numerous, separated by 7-9cm. Leaves dimorphic, glabrous; sterile leaves oblong, 7-8cm × 4-7cm, odd pinnae, pinna small, cordate or ovate-triangular, 1.5-3cm long, apex obtuse, base cordate, margin short serrated. Fertile leaves same shape, 8-10cm × 4-6cm, pinnule triangle or ovate-triangular. Fertile spike bar, 3-6mm long, glabrous, each spike sporangium 5-8, arranged at the leaf margin.

Distributed in Hunan(Tongdao, Jianghua, Yizhang), Fujian, Guangdong, Guangxi, Hainan, Hongkong, Jiangxi, Taiwan and South Yunnan, also in other tropical regions in Asian. Grows on mountain slopes and in valley shrubs at low altitude.

金毛狗　*Cibotium barometz* (Linn.) J. Sm.

植株树状，高达3m。根状茎粗壮，匍匐，密被金黄色长茸毛，形如金毛狗头，顶端有叶丛生。叶柄粗，长达1m以上；叶片革质，阔卵状三角形，2回羽状，长1.5～3m，上面深绿色，下表面被白霜，除中脉被毛外，两面均无毛；末回裂片镰状披针形，长1～1.4cm，宽约3mm，尖头，边缘有浅锯齿，侧脉不分叉，或在不育裂片上为二叉。孢子囊群生于小脉顶端，每裂片1～5对；囊群盖2瓣裂，形如蚌壳。

产莽山、江华、江永、通道、汝城、桂东、东安、保靖。生沟谷或林下阴湿处。分布于浙江南部、江西、福建、台湾、广东、广西、香港、海南、台湾、贵州、四川、云南南部、西藏。亚洲热带其他地区亦产。

Plant like tree, up to 3m tall. Rhizomes stout, prostrate, densely covered with golden long hairs, shape like a golden nugget, apical leaves clustered. Petiole thick, up to 1 m long or more; leaf blade leathery, broadly ovate-triangular, bipinnate, 1.5-3m long, dark green adaxially, pruinose abaxially, except midrib with hairy, glabrous on both surfaces; ultimate lobes sickle lanceolate, ca.1-1.4cm × 3mm, apex acute, margin shallowly dentate, lateral veins not bifurcated, or binary in sterile lobes. Sori borne at the top of small veins, each lobe 1-5 pairs; indusium 2-valved, like shell.

Distributed in Hunan(Mangshan, Jianghua, Jiangyong, Tongdao, Rucheng, Guidong, Dongan, Baojing), South Zhejiang, South Jiangxi, Fujian, Taiwan, Guangdong, Guangxi, Hongkong, Hainan, Taiwan, Guizhou, Sichuan, South Yunnan and Tibet, also in other tropical regions in Asia. Grows in gullies or moist places in forests.

植株灌木状，高0.6～1.4m。主干短而横卧。叶簇生；叶柄长30～90cm，红褐色，基部被金黄色鳞片；鳞片线形，淡棕色，光亮，边缘具疏长刚毛；叶片披针形，长35～50cm，2～3回羽状；羽片12～16对，互生，具短柄，长圆形；小羽片长7～8cm，宽1.6～1.8cm，先端短渐尖，无柄，深羽裂，基部1～2裂片分离；裂片边缘具粗齿；小脉单一；羽轴红棕色，疏生线形鳞片；小羽片主脉及裂片中脉背面密生泡状小鳞片。孢子囊群圆形；囊群盖缺。

产江华、江永。生山坡林下、沟谷边。分布于重庆、福建、广东、广西、贵州、香港、江西、四川、台湾、云南、浙江。日本南部亦产。

Plant like shrub, 0.6-1.4m tall. Trunk short and recumbent. Leaves tufted; petioles 30-90cm long, rufous, base with golden scales; scales linear, light brown, shiny, margin with sparse long setose; leaf blade lanceolate, 35-50cm long, bipinnate to tripinnate; pinnae 12-16 pairs, alternate, with short petiole, oblong; pinnules 7-8cm × 1.6-1.8cm, apex shortly acuminate, sessile, pinnatipartite, 1-2 lobes separated at the base; lobes margin coarse dentate; small veins single; rachis reddish-brown, sparsely linear scales; main vein of pinnule and midvein of lobes densely vesicular small scales abaxially. Sori round; indusium missing.

Distributed in Hunan(Jianghua, Jiangyong), Chongqing, Fujian, Guangdong, Guangxi, Guizhou, Jiangxi, Sichuan, Taiwan, Yunnan and ZheJiang, also in South Japan. Grows in hillside forests or in gullies.

桫椤 *Alsophila spinulosa* (Wall. ex Hook.) Tryon

植株灌木状，高达6m。叶螺旋状排列于茎顶端；茎段端、拳卷叶、叶柄基部密被鳞片和鳞毛，鳞片暗棕色，有光泽；叶柄长30～50cm，通常棕色，连同叶轴和羽轴有刺状突起；叶片大，长矩圆形，长1～2m，宽0.4～0.5m，3回羽状深裂；羽片17～20对，互生；小羽片18～20对；裂片18～20对，镰状披针形，边缘有锯齿；小脉常2叉；羽轴和中脉下面被灰白色小鳞片。孢子囊群着生于侧脉分叉处，靠近中脉；囊群盖球形，外侧开裂，熟时反折覆盖于主脉上。

产东安。生林下或溪边。分布于江西、四川、台湾、西藏、云南、重庆、福建、广东、广西、贵州、海南、香港。茎含淀粉，可供食用。优良庭园观赏树种。

Plant like shrub, up to 6m tall. Leaves spirally arranged at the top of stem; top of stem segments, circinate leaves, base of petioles densely scaly and scaly hairs, scales dark brown, shiny; petioles 30-50cm long, usually brown, petioles, rachis and rachis with spiked protuberances; leaf blade large, long oblong, 1-2m × 0.4-0.5m, tripinnate; pinnae 17-20 pairs, alternate; pinnules 18-20 pairs; lobes 18-20 pairs, sickle lanceolate, margin serrate; small veins often 2 fork; rachis and midrib abaxially with grayish white small scales. Sori borne on bifurcated place of lateral veins, near midrib; indusium globose, external cracking, reflexed and covered in main veins when mature.

Distributed in Hunan (Dongan), Jiangxi, Sichuan, Taiwan, Tibet, Yunnan, Chongqing, Fujian, Guangdong, Guangxi, Guizhou, Hainan, Hongkong. Grows in forests or steamsides. The stems contain starch, edible. It is an excellent ornamental tree.

小黑桫椤（华南黑桫椤） *Alsophila metteniana* (Hance) Tagawa

植株灌木状，高1.5～2m。根状茎短而斜伸，密生黑棕色鳞片。叶密生，叶柄长70～90cm，棕黑色，基部密生宿存的鳞片，向上渐稀疏；叶轴及羽轴腹面密生短毛；叶片狭卵形，长80～130cm，3回羽状分裂；羽片15～20对，互生，具柄，卵状披针形，2回羽状分裂；小羽片15～20对，互生或近对生，长6～9cm，宽1.5～2.2cm；裂片狭长，先端具小圆齿；小脉单一；羽轴背面及裂片主脉背面生小鳞片。孢子囊群圆形，无盖。

产通道。散生低海拔潮湿沟谷林下。分布于四川、云南、福建、江西、台湾。日本亦产。

Plant like shrub, 1.5-2m tall. Rhizomes short, and obliquely extending, densely black-brown scales. Leaves dense, petioles 70-90cm long, brown-black, dense persistent scales at the base, upwards gradually sparse; rachis and pinna rachis adaxially dense shorthair; leaves narrowly ovate, 80-130cm long, tripinnatifid; pinnae 15-20 pairs, alternate, petiolate, ovate-lanceolate, bipinnatifid; pinnules 15-20 pairs, alternate or subopposite, 6-9cm × 1.5-2.2cm; lobes narrowly long, apex with small crenate; small veins single; main vein and rachis abaxially with small scales. Sori round, no indusium.

Distributed in Hunan(Tongdao), Sichuan, Yunnan, Fujian, Jiangxi and Taiwan, also in Japan. Grows in moist ravine forests at low altitude.

半边旗　*Pteris semipinnata* L.

多年生草本，高35～80cm。根状茎横走，顶端及叶柄基部被钻形鳞片。叶2型，近簇生，草质，全株近光滑；叶柄栗色至深栗色，四棱形；能育叶片长圆形或长圆状披针形，长20～40cm，2回半边羽状深裂；羽片三角形或半三角形，长尾尖，上侧全缘，下侧羽深裂；裂片宽3～6mm，基部的最长，向上逐渐变短，边缘仅不育叶的顶部有尖锯齿；不育叶同形，全有锯齿。脉明显，外倾，分叉或2回羽状分叉，小脉伸到锯齿基部。

产通道、宜章。生林下或石上。分布于江西、澳门、四川、台湾、云南、浙江、重庆、福建、广东、广西、贵州、海南、香港。亚洲热带其他地区亦产。

Perennial herbs, 35-80cm tall. Rhizomes horizontal, top and petiole base subulate scales. Leaves dimorphic, nearly fascicled, herbaceous, nearly smooth throughout; petioles maroon to dark maroon, 4-angled; fertile leaves oblong or oblong-lanceolate, 20-40cm long, bipinnatiparted at one side; pinnae triangular or semi-triangular, long caudate-acuminate, upper side margin entire, lower side pinnatipartite; lobes 3-6mm wide, longest at base, upwards gradually shorter, margin sharp dentate only at the top of sterile leaves; sterile leaves homomorphic, margin dentate. Veins conspicuous, decumbent, forked or bipinnate forked, veinlets straight direct to serrated base.

Distributed in Hunan(Tongdao, Yizhang), Jiangxi, Macao, Sichuan, Yunnan, ZheJiang, Chongqing, Fujian, Guangdong, Guangxi, Guizhou, Hainan and Hongkong, also in other tropical regions in Asia. Grows in forests or on rocks.

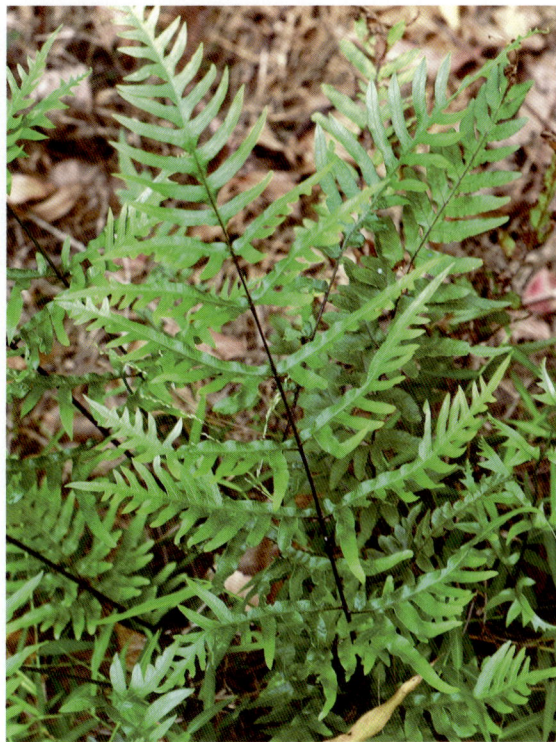

毛 蕨　*Cyclosorus interruptus* (Willd.) H. Ito

多年生草本，高达1.3m。根状茎横走，黑色，顶部被少数鳞片。叶柄长约70cm，基部黑褐色，向上深禾秆色，近光滑；叶卵状披针形或椭圆披针形，长约60cm，先端尾状渐尖，2回羽状；叶脉下面明显，每裂片侧脉8～10对，基部一对斜伸，其上侧1脉出自主脉基部，下侧1脉出自羽轴，顶端交结成三角形网眼；第2对侧脉斜达膜质联线；叶近革质，上面无毛，下面沿羽轴疏被卵形鳞片。孢子囊群圆形，生侧脉中部，每裂片5～9对，下部1～2对不育。

产江华、江永。生于海拔150～350m的溪边。分布于福建、广东、广西、海南、香港、江西、澳门、台湾。热带和亚热带其他地区亦产。

Perennial herbs, up to 1.3m tall. Rhizomes horizontal, black, top covered with few scales. Petioles ca.70cm long, base dark brown, upwards dark straw color, nearly smooth; leaves ovate-lanceolate or elliptic-lanceolate, ca.60cm long, apex caudate acuminate, bipinnate; conspicuous veins abaxially, each lobe lateral veins 8-10 pairs, base 1 pairs obliquely splay, upper side 1 vein from main vein base, lower side 1 vein from rachis, apex cross to form triangle mesh; Second lateral veins inclined to membranous on-line; leaves subleathery, glabrous adaxially, along rachis sparsely ovate scales abaxially. Sori round, borne on lateral veins middle, each lobe 5-9 pairs, lower 1-2 pairs infertile.

Distributed in Hunan (Jianghua, Jiangyong), Fujian, Guangdong, Guangxi, Hainan, Hongkong, Jiangxi, Macau and Taiwan, also in other tropical and subtropical regions. Grows in streamsides at alt. 150-350m.

多年生草本，高0.8~1.1m。根状茎短粗，密被红棕色鳞片。叶近簇生；叶柄粗壮，长20~50cm，褐色，基部被鳞片；叶片长卵形或椭圆形，长20~60cm，2回羽状深裂；羽片互生，线状披针形，先端长渐尖，基部极不对称，1回深羽裂；叶厚纸质，无毛，裂片上面中脉两侧生小珠芽，珠芽萌发幼叶，落地后长成新植株。孢子囊群近新月形；囊群盖长肾形，革质，宿存。

产汝城、宁远。生于海拔150~800m林下溪沟边或路边。分布于江西、浙江、福建、广东、广西、台湾。小珠芽生于叶片，姿态极优美，适于盆栽观赏。

Perennial herbs, 0.8-1.1m tall. Rhizomes short and thick, densely reddish-brown scales. Leaves nearly clustered; Petioles stout, 20-50cm long, brown, base scales; leaves long ovate or elliptic, 20-60cm long, bipinnatiparted; pinnae alternate,linear-lanceolate, apex long acuminate, base very asymmetric, pinnatiparted; leaves thickly papery, glabrous, both sides of lobe midvein adaxially with bulbils, bulbil sprouting young leaves, after fall to the ground, grow into new plants. Sori nearly crescent-shaped; indusium long kidney-shaped, leathery, persistent.

Distributed in Hunan(Rucheng, Ningyuan), Jiangxi, Zhejiang, Fujian, Guangdong, Guangxi and Taiwan. Grows in streamsides under forests or along roadsides at alt. 150-800m. Small bulbils grows on the leaves, the shape is very beautiful and suitable for potted ornamental.

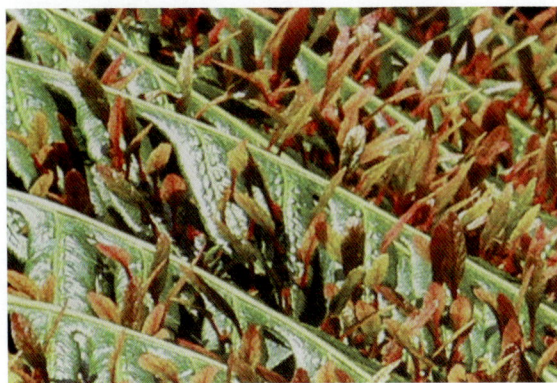

海南五针松　　　*Pinus fenzeliana* H.-M.

常绿乔木。小枝淡褐色。冬芽红褐色。叶5针1束，细长柔软，通常长10~18cm；叶鞘早落；叶内具1条维管束。雄球花卵圆形，多数聚生于新枝下部成穗状，长约3cm。球果长卵圆形或椭圆状卵圆形，单生或2~4簇生于小枝基部；种鳞鳞盾边缘明显向外反卷，种子顶端具短翅。花期4月，种子翌年10~11月成熟。

产新宁、通道、东安。常生于石灰岩山地。分布于贵州、四川、广东、海南、广西。木材质软，纹理细腻，可作建筑用材，亦可提取松脂。

Evergreen trees. Twigs pale brown. Winter buds rufous. Needles 5 per bundle, slender and soft, usually 10-18cm long; sheaths caducous; leaves with 1 vascular bundle. Male cones ovate-orbicular, most clustered at lower part of new branches form spikes, ca.3cm long. Cone long-ovoid or elliptic-ovoid, solitary or 2-4 clustered at the base of branchlets; Seed scales shield edge obviously reflexed, seeds top with short wings. Fl. Apr, seed maturity Oct-Nov of 2nd year.

Distributed in Hunan(Xinning, Tongdao, Dongan), Guizhou, Sichuan, Guangdong, Hainan and Guangxi. Usually grows in limestone mountain. Wood material is soft, delicate texture, can be used for construction timber, and also can be extracted turpentine.

竹 柏　　*Nageia nagi* (Thunb.) O. Kuntze

常绿乔木。叶对生，革质，长卵形至卵状披针形，具平行细脉，无中脉，上面深绿色，有光泽。雄球花穗状柱形，单生叶腋，常呈分枝状；雌球花单生叶腋，稀成对腋生。种子圆球形，径1.2～1.5cm，成熟时假种皮暗紫色，被白粉。花期2～4月，种子10月成熟。

产城步、新宁、通道、江华、江永。生山谷、溪边林中。分布于浙江、福建、广东、广西、海南、江西、四川、台湾。树姿优雅，系优良园林绿化树种。湖南南岭地区稀见种。

Evergreen trees. Leaves opposite, leathery, long ovate to ovate-lanceolate, with parallel veins, without midvein, dark green adaxially, shiny. Male cones spicate column, solitary and axillary, usually branched; female cone solitary and axillary, rarely paired axillary. Seed globose, 1.2-1.5cm in diam., epimatium dark purple when mature, pruinose. Fl. Feb-Apr, seed maturity in Oct.

Distributed in Hunan(Chengbu, Xinning, Tongdao, Jianghua, Jiangyong), Zhejiang, Fujian, Guangdong, Guangxi, Hainan, Jiangxi, Sichuan and Taiwan. Grows in valley and streamsides in forests. It is an excellent landscaping tree because of its elegant posture. It is also a rare species in Nanling region of Hunan.

百日青　*Podocarpus neriifolius* D. Don

常绿乔木。叶螺旋状互生，披针形，厚革质，常微弯，长7~15cm，先端长渐尖，上面中脉隆起。雄球花穗状，单生或2~3簇生。种子卵圆形，长8~16mm，顶端圆或钝，熟时假种皮紫红色。花期5月，种子10~11月成熟。

产江华、江永、通道、新宁、桂东、宜章、汝城。生山地潮湿林中。分布于浙江南部、福建、广东、广西、贵州、云南、四川、西藏。喜马拉雅地区、东南亚亦产。

Evergreen trees. Leaves spirally alternate, lanceolate, thickly leathery, usually slightly curved, 7-15cm long, apex long acuminate, midvein raised adaxially. Male cones spicate, solitary or 2-3fascicled. Seed ovate-orbicular, 8-16mm long, apex rounded or obtuse, epimatium purple-red when mature. Fl. May, seed maturity Oct-Nov.

Distributed in Hunan(Jianghua, Jiangyong, Tongdao, Xinning, Guidong, Yizhang, Rucheng), South Zhejiang, Fujian, Guangdong, Guangxi, Guizhou, Yunnan, Sichuan and Tibet, also in Himalaya regions and Southeast Asia. Grows in humid forests in mountain land.

毛桃木莲　　*Manglietia moto* Dandy

常绿乔木。芽、幼枝、嫩叶、托叶、叶背面、叶柄、花梗及果梗均密被锈褐色茸毛，小枝具托叶环痕。叶革质，长圆形或倒卵状椭圆形，长12～25cm，先端短钝尖或渐尖，基部楔形或宽楔形，上面无毛，下面密被锈褐色茸毛，沿中脉更密。花白色，单生枝顶。聚合果卵形或长卵形。花期5～6月，果期10～12月。

产宜章、资兴。生山地沟谷密林中。分布于福建、广东、广西。枝繁叶茂，树冠宽广，花果艳丽，为优良的庭园绿化树种。

Evergreen trees. Buds, young twigs, young leaves, stipules, lower surface of leaves, petioles, fruiting pedicels densely rusty-brown tomentose, twigs with annular stipular scars. Leaves leathery, oblong or obovate-elliptic, 12-25cm long, apex short obtusely pointed or acuminate, base cuneate or broadly cuneate, glabrous adaxially, densely rusty-brown villous abaxially, more densely so along midvein. Flowers white, solitary and terminal. Aggregate fruit ovoid or long ovate. Fl. May-Jun, fr. Oct-Dec.

Distributed in Hunan(Yizhang, Zixing), Fujian, Guangdong and Guangxi. Grows in dense forests in mountain ravine. It is an excellent garden plant because of its dense twigs and leaves, broad crown, and attractive flowers and fruits.

醉香含笑（火力楠） *Michelia macclurei* Dandy

常绿乔木。芽、幼枝、叶柄、托叶及花梗密被紧贴红褐色短柔毛，小枝具托叶环痕。叶革质，倒卵形或椭圆状倒卵形，长7～14cm，先端急尖，基部楔形，下面被灰色毛。花黄白色，单生叶腋，花被片9；花梗密被红褐色柔毛。花期3～4月，果期9～11月。

产通道。生山地密林中。分布于贵州、广西、福建、浙江南部、广东。越南北部亦产。花可提取香精油。树形美观，花芳香，为优良庭院观赏树种及行道树。

Evergreen trees. Buds, young twigs, petioles, stipules and pedicels densely rufous appressed pubescent, twigs with annular stipular scars. Leaves leathery, obovate or elliptic-obovate, 7-14cm long, apex acute, base cuneate, gray pubescent abaxially. Flowers yellow-white, solitary and axillary, tepals 9; pedicels densely rufous pubescent. Fl. Mar-Apr, fr. Sep-Nov.

Distributed in Hunan(Tongdao), Guizhou, Guangxi, Fujian, South Zhejiang and Guangdong, also in North Vietnam. Grows in dense forests in mountain land. Flowers can be extracted essential oils. It is an excellent garden ornamental tree and street tree because of its beautiful tree-shape and fragrant flowers.

观光木（香花木） *Tsoongiodendron odorum* Chun

常绿乔木。芽、小枝、叶柄、叶上面中脉、叶背及花梗密被黄棕色糙伏毛，小枝具托叶环痕。叶厚纸质，倒卵状椭圆形，先端骤尖，基部楔形。花两性，单生叶腋，芳香，花梗长约6mm，花被片9，淡黄色，具红色小斑点。聚合果硕大，木质，椭圆形。花期3～4月，果期10～11月。

产新宁、通道、江华、宜章、江永、汝城。生低山沟谷阔叶林中。分布于福建、江西南部、广东、海南、广西、云南东南部、贵州。越南北部亦产。为我国特有孑遗种和稀有种，国家Ⅱ保护植物。

Evergreen trees. Buds, twigs, petioles, midribs adaxially, lower surface of leaves and pedicels densely yellowish-brown strigose, twigs with annular stipular scars. Leaves thickly papery, obovate-elliptic, apex cuspidate, base cuneate. Flowers bisexual, solitary and axillary, fragrant, pedicels ca.6mm long, tepals 9, pale yellow, with red punctate. Aggregate fruit large, woody, elliptic. Fl. Mar-Apr, fr. Oct-Nov.

Distributed in Hunan(Xinning,Tongdao, Jianghua, Yizhang, Jiangyong, Rucheng), Fujian, South Jiangxi, Guangdong, Hainan, Guangxi, Southeast Yunnan and Guizhou, also in North Vietnam. Grows in broad-leaved forests in low mountain ravine. It is a unique relict species and rare species of China, Ⅱ national protected plants.

红花八角　*Illicium dunnianum* Tutch.

常绿灌木。幼枝纤细。叶集生近枝顶，3～8片簇生，或假轮生，薄革质，狭披针形或狭倒披针形，长5～12cm，先端急尾状渐尖或渐尖，基部渐狭，下延至叶柄成明显狭翅，中脉在叶上面稍凹下。花单生或2～3簇生；花梗纤细，长1～3.5cm；花被片12～20，粉红色至红色、紫红色，椭圆形到近圆形；心皮8～13。果较小，径1.5～3cm，蓇葖通常7～8枚，少13枚，顶端有明显钻形尖头，长3～5mm，略弯曲。花期3～7月，果期7～10月。

产通道。生山谷溪边、密林阴湿处。分布于福建、广东、广西、贵州。

Evergreen shrubs. Young twigs slender. Leaves crowded at the nearly top of branchlets, 3-8-fascicled, or pseudoverticillate, thinly leathery, narrowly lanceolate or narrowly oblanceolate, 5-12cm long, apexcaudate-acuminate or acuminate, base attenuate, decurrent to petiole form obvious narrow wings, midrib slightly impressed adaxially. Flowers solitary or 2-3-fascicled; pedicels slender, 1-3.5cm long; tepals 12-20, pink to red, purple-red, elliptic to suborbicular; carpels 8-13. Fruit small, 1.5-3cm in diam., follicles usually 7-8, rarely 13, top with obvious subulate cusp, 3-5mm long, slightly curved. Fl. Mar-Jul, fr. Jul-Oct.

Distributed in Hunan(Tongdao), Fujian, Guangdong, Guangxi and Guizhou. Grows in valley streamsides and in moist places in dense forests.

常绿乔木。幼枝棕色或紫色，后变灰色，具皮孔。叶3～6片排成不整齐的假轮生，近革质，长圆状披针形或倒披针形，长10～20cm，先端渐尖，尖头长8～20mm，基部楔形，中脉在叶上面轻微凹陷。花近顶生或腋生，单生或2～4朵簇生；花梗长1.8～6cm；花被片15～21，红色；心皮11～14枚，极少9枚。果径较大，径4～4.5cm，蓇葖10～14枚，顶端有明显钻形尖头，长3～7mm。花期4～6月，果期7～10月。

产城步、道县、江华、新宁。生山地阔叶林中。分布于广西、广东、云南、贵州。

Evergreen trees. Young twigs brown or purple, later becoming gray, lenticellate. Leaves 3-6 arranged in intoirregular pseudoverticillate, subleathery, oblong-lanceolate or oblanceolate, 10-20cm long, apex acuminate, apex 8-20mm long, base cuneate, midvein slightly impressed adaxially. Flowers nearly terminal or axillary, solitary or 2-4-fascicled; pedicels 1.8-6cm long; tepals 15-21, red; carpels 11-14, rarely 9. Fruit larger, 4-4.5cm in diam., follicles 10-14, top with obvious subulate cusp, 3-7mm long. Fl. Apr-Jun, fr. Jul-Oct.

Distributed in Hunan(Chengbu, Daoxian, Jianghua, Xingning), Guangxi, Guangdong, Yunnan and Guizhou. Grows in broad-leaved forests in mountain land.

攀援灌木。小枝被黄色粗毛。叶革质，椭圆状长圆形至长圆形，先端急尖或钝，基部近圆形或少偏斜，侧脉8~10对，两面凸起，无毛或仅中脉下面被短柔毛。花单生于钩状总花梗上，花梗略长于总花梗；花瓣卵状披针形，长10~18mm；心皮卵状长圆形，柱头短棍棒状。果椭圆形，干时黑色，先端近圆形。花期3~5月，果期5~12月。

产宜章、通道、江永。生山地密林下或潮湿山谷。分布于广东、广西、云南、贵州。越南亦产。

Climbing shrubs. Branchlets yellowish hirsute. Leaves leathery, elliptic-oblong to oblong, apex acute or obtuse, base subrounded or slightly oblique; lateral veins 8-10 pairs, rasied on both surfaces, glabrous or only midvein abaxially puberulent. Flower solitary in hooked peduncle, pedicels slightly longer than peduncle; petals ovate-lanceolate, long 1-1.8cm; carpels ovate-oblong, stigma shortly clavate. Fruit ellipsoid, black when dry, apex subrounded. Fl. Mar-May, fr. May-Dec.

Distributed in Hunan(Yizhang, Tongdao, Jiangyong), Guangdong, Guangxi, Yunnan and Guizhou, also in Vietnam. Grows in dense forests in mountain land or wet valleys.

瓜馥木 *Fissistigma oldhamii* (Hemsl.) Merr.

攀援木质藤本。小枝被黄褐色柔毛。叶革质，倒卵状椭圆形或长圆形，先端圆或微凹，基部宽楔形至圆形，背面被黄褐色短柔毛，侧脉在叶上平坦。花单生或2～3聚生密伞花序。果球形，密被棕黄色柔毛。花期4～9月，果期7月至翌年2月。

产城步、通道、新宁、道县、江华、江永、宜章、汝城。生低山沟谷或林缘。分布于广西、广东、云南、福建、江西、浙江、台湾。越南亦产。

Climbing woody vines. Twigs yellow-brown pilose. Leaves leathery, obovate-elliptic or oblong, apex rounded or retuse, base broadly cuneate to rounded, yellow-brown pubescent abaxially, lateral veins compressed adaxially. Flowers solitary or 2-3 aggregated in cymose inflorescences. Fruit globose, densely brown villous. Fl. Apr-Sep, fr. Jul to Feb of 2nd year.

Distributed in Hunan(Chengbu, Tongdao, Xinning, Daoxian, Jianghua, Jiangyong, Yizhang, Rucheng), Guangxi, Guangdong, Yunnan, Fujian, jiangxi, Zhejiang and Taiwan, also in Vietnam. Grows in ravine or forest margins in low mountain.

香港瓜馥木　*Fissistigma uonicum* (Dunn) Merr.

攀援木质藤本。枝无毛。叶纸质，长圆形，长4～20cm，先端急尖，基部圆形或楔形，侧脉在叶上稍凸起。花序腋生，具1～2花，花序梗长1～3mm；花黄色，花梗长约2cm；外轮花瓣比内轮花瓣长，无毛，卵状三角形。果球形，熟时黑色，被短柔毛。花期3～6月，果期6～12月。

产通道、城步、江永、江华、宜章、汝城。生低山沟谷的林缘或灌丛。分布于福建、广东、贵州、海南。印度尼西亚亦产。果甜可食。

Climbing woody vines. Branches glabrous. Leaves papery, oblong, 4-20cm long, apex acute, base rounded or cuneate, lateral veins slightly raised adaxially. Inflorescences axillary, with 1-2 flowers, peduncle 1–3 mm long; flowers yellow, pedicels ca.2cm long; outer petals longer than inner petals, glabrous, ovate-triangular. Fruit globose, black when mature, pubescent. Fl. Mar-Jun, fr. Jun-Dec.

Distributed in Hunan (Tongdao, Chengbu, Jiangyong, Jianghua, Yizhang, Rucheng), Fujian, Guangdong, Guizhou and Hainan, also in Indonesia. Grows in forest margins or thickets in low mountain ravine. Fruit is sweet and edible.

野独活 *Miliusa chunii* W. T. Wang

灌木。叶膜质，椭圆形或椭圆状长圆形，先端渐尖或短渐尖，基部偏斜，宽楔形或圆形，全缘或呈波状，侧脉10～12对，在近叶缘处连结。花红色，单生叶腋，花梗丝状，长4～6.5cm。小果圆球形，熟时紫黑色，无毛，具长约4～7.5cm的总果梗。花期5～7月，果期8～12月。

产江永。生石灰岩山地或山谷灌丛。分布于广东、广西、云南、海南。越南亦产。

Shrubs. Leaves membranous, elliptic or elliptic-oblong, apex acuminate or short acuminate, base oblique, broadly cuneate or rounded, margin entire or wavy, lateral veins 10-12 pairs, anastomosing near margin. Flowers red, solitary and axillary, pedicels filiform, 4-6.5cm long. Small fruit globose, dark purple when mature, glabrous, fruiting pedicel ca.4-7.5cm long. Fl. May-Jul, fr. Aug-Dec.

Distributed in Hunan (Jiangyong), Guangdong, Guangxi, Yunnan and Hainan, also in Vietnam. Grows in limestone mountain and valley thickets .

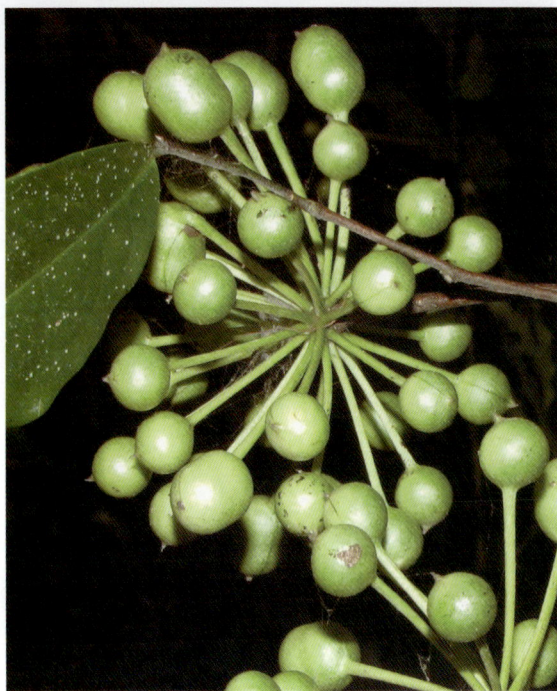

常绿乔木。顶芽小，卵球形，密被黄褐色短柔毛。叶互生或近对生，革质，椭圆形至长圆状椭圆形，长5.5～13.5cm，先端急尖或短渐尖，中脉在上面凹陷，下面凸起，小脉纤细，密网状，两面稍凸起，干后略成蜂窝小穴；叶柄长5～14mm。圆锥花序腋生，长3～8cm。果椭圆形，长1.5～2.3cm，熟时紫黑色，被小瘤状凸起。花期5～8月，果期7～12月。

产汝城。生山坡湿润常绿阔叶林或灌丛中。分布于广东、广西、云南、台湾、贵州西南部。越南北部亦产。湖南省新记录种。

Evergreen trees. Terminal buds small, ovoid, densely yellow-brown pubescent. Leaves alternate or subopposite, leathery, elliptic to oblong-elliptic, 5.5-13.5cm long, apex acute or short acuminate, midvein impressed adaxially, raised abaxially, small veins slender, densely reticulate, slightly elevated on both surfaces, after dry slightly foveolate; petioles 5-14mm long. Panicles axillary, 3-8cm long. Fruit elliptic, 1.5-2.3cm long, purple-black when mature, verruculose. Fl. May-Aug, fr. Jul-Dec.

Distributed in Hunan(Rucheng), Guangdong, Guangxi, Yunnan, Taiwan and Southwest Guizhou, also in North Vietnam. Grows in humid evergreen broad-leaved forests on mountain slopes or thickets. A new recorded species in Hunan province.

寄生缠绕草本。茎线形，绿色或绿褐色，无毛或稍有毛。叶退化为微小鳞片。穗状花序，长2～5cm；花极小，两性，白色，长不到2mm，无花梗；花被片6，宿存，成2轮，外轮3枚小，圆形，有绿色毛，内轮3枚大，卵形；能育雄蕊9，成3轮。果实小，卵球形，径约7mm，包藏于肉质果托内。

产江永、江华。生山坡灌丛或疏林中。分布于福建、广东、广西、贵州、海南、江西、台湾、云南、浙江。全球其他热带地区亦产。本种对寄主有害。

Parasitic twining herbs. Stem linear, green or green-brown, glabrous or slightly hairy. Leaves reduced to minute scales. Spikes, 2-5cm long; flowers very small, bisexual, white, less than 2 mm long, sessile; perianth segments 6, persistent, in 2 series, outer 3 small, orbicular, with green hair, inner 3 larger, ovate; fertile stamens 9, in 3 series. Fruit small, ovoid, ca.7mm in diam., wrapped in fleshy fruit receptacle.

Distributed in Hunan (Jianghua, Jiangyong), Fujian, Guangdong, Guangxi, Guizhou, Hainan, Jiangxi, Taiwan, Yunnan and Zhejiang, also in other tropical regions of world. Grows in thickets or sparse forests on mountain slopes. This pantropical species is harmful to its host plant.

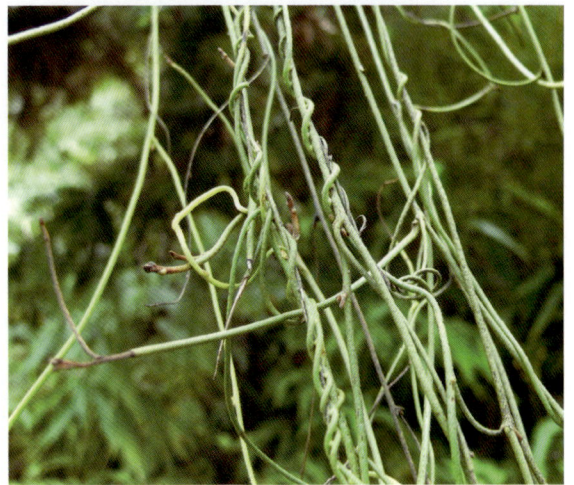

华南桂　*Cinnamomum austro-sinense* H. T. Chang

常绿小乔木。叶近对生，厚革质，椭圆形或长椭圆形，长11～22cm，背面密被贴伏灰褐色微柔毛，基出3脉或近离基3出脉，与中脉在上面凸起或微凸起；叶柄长1.5～2cm。花黄绿色，排成圆锥花序，序轴及花梗被灰色绢毛。果椭圆形，果托浅杯状，托口平截。花期6～8月，果期8～10月。

产通道、城步、绥宁、道县、江华、宜章、汝城。生山坡或沟谷阔叶林中。分布于福建、广东、广西、江西、浙江南部。树皮芳香，可代桂皮用。叶研粉，可作熏香原料。

Evergeen small trees. Leaves subopposite, thickly leathery, elliptic or long elliptic, 11-22cm long, densely appressed gray brown puberulent abaxially, basal veins 3 or nearly triplinerved, and midrib raised or slightly raised adaxially; petioles 1.5-2cm long. Flowers yellowish-green, arranged in panicles, inflorescence rachis and pedicels grayish sericeous. Fruit elliptic, fruit receptacle shallowly cupular, receptacle mouth flatted truncate. Fl. Jun-Aug, fr. Aug-Oct.

Distributed in Hunan (Tongdao, Chengbu, Suining, Daoxian, Jianghua, Yizhang, Rucheng), Fujian, Guangdong, Guangxi, Jiangxi and South Zhejiang. Grows in broad-leaved forests on mountain slopes or ravines. Aromatic bark can be used as a substitute for cinnamon. Leaves can be used as incense raw materials by milling into powder.

辣汁树　*Cinnamomum tsangii* Merr.

常绿乔木。叶近对生，薄革质，披针形或长圆状披针形，长5～10cm，幼时两面被银白色绢毛，老叶背面被浅褐色绢毛，近基出3脉，与中脉在上面凸起或微凸起；叶柄长5～12mm。花绿白色，排成聚伞状。果椭圆形，长1.5cm；果托浅杯状。花期10月。

产新宁、通道、宜章。生沟谷灌丛或阔叶林中。分布于福建、广东、广西、四川、江西南部。

Evergreen trees. Leaves subopposite, thinly leathery, lanceolate or oblong-lanceolate, 5-10cm long, silver white sericeous on both surfaces of young Leaves, old leaves abaxially light brown sericeous, near basal veins 3, and with midribs raised or slightly raised adaxially; petioles 5-12mm long. Flower green-white, arranged in cymose. Fruit elliptic, 1.5cm long; fruit receptacle shallowly cupular. Fl. Oct.

Distributed in Hunan (Xinning, Tongdao, Yizhang), Fujian, Guangdong, Guangxi, Sichuan and South Jiangxi. Grows in shrubs or broad-leaved forests in ravines.

常绿乔木。叶近对生，硬革质，椭圆形，长4.5~9.5cm，上面光亮，下面无毛，偶被白粉，基出3大脉强劲几达叶尖，与中脉在上面微凹；叶柄长6~8mm。圆锥花序，具疏花，花序梗与叶近等长。花期7月。

产宜章。生山地疏林。分布于广东、广西。

Evergreen trees. Leaves subopposite, hard leathery, elliptic, 4.5-9.5cm long, shiny adaxially, glabrous abaxially, rarely pruinose, basal three veins robustly almost reach leaf apex, and with midribs slightly impressed adaxially; petioles 6-8mm long. Panicles, laxly flowered, peduncle ca. as long as leaves. Fl. Jul.

Distributed in Hunan (Yizhang), Guangdong and Guangxi. Grows in sparse forests in mountain land.

常绿乔木。叶革质，互生或近对生，长椭圆形，长7～11cm，先端突渐尖或短尖，基部阔楔形，上面光亮，下面苍白色，具离基3出脉，中脉在上面凹陷，下面凸起。圆锥花序腋生及顶生，花序总梗被黄褐色短柔毛；花淡黄色，花被筒钟状。果球形或扁球形，全部包藏于扩大的花被筒中，熟时紫黑色，外具12～15条纵棱。花期4～5月，果期8～12月。

产江永、江华。生沟谷常绿阔叶林中。分布于广东、广西、江西、福建、浙江南部、台湾。越南亦产。

Evergreen trees. Leaves leathery, alternate or subopposite, long elliptic, 7-11cm long, apex abruptly acuminate or acute, base broadly cuneate, shiny adaxially, pale abaxially, triplinerved, midvein impressed adaxially, raised abaxially. Panicles axillary and terminal, inflorescence peduncle yellow-brown pubescent; flowers pale yellow, perianth tube campanulate. Fruit ovoid or oblate, all wrapped in enlarged perianth tube, violet-black when mature, outside with 12-15 longitudinal ridges. Fl. Apr-May, fr. Aug-Dec.

Distributed in Hunan (Jiangyong, Jianghua), Guangdong, Guangxi, Jiangxi, Fujian, South Zhejiang and Taiwan, also in Vietnam. Grows in evergreen broad-leaved forests in ravines.

广东润楠 *Machilus kwangtungensis* Yang

常绿乔木。幼枝密被锈色茸毛，老枝无毛。叶革质，长椭圆形或倒披针形，长6～15cm，先端渐尖，基部渐狭，上面无毛，下面被贴伏短柔毛，中脉上面凹陷，下面凸起，侧脉10～12对，纤细，两面不明显；叶柄长8～15mm，被短柔毛。圆锥花序生新枝下端，长5～10.5cm，被灰黄色短柔毛；花被裂片近等长，长圆形，长约5mm，两面均被短柔毛。果近球形，略扁，径8～9mm，熟时黑色。花期3～4月，果期5～7月。

产宜章、通道。生山谷阔叶林、石灰岩山地。分布于广东、广西、贵州南部。

Evergreen trees. Young branchlets densely rusty tomentose, old branches glabrous. Leaves oblong or oblanceolate, 6-15cm long, leathery, apex acuminate, base attenuate, glabrous adaxially, appressed pubescent abaxially, midvein impressed adaxially, raised abaxially, lateral veins 10-12 pairs, slender, inconspicuous on both surfaces; petioles 8-15mm long, pubescent. Panicles arising from lower part of current year branchlet, 5-10.5cm long, grayish yellow pubescent, perianth lobes subequal, oblong, ca.5mm, pubescent on both surfaces. Fruit subglobose, slightly compressed, 8-9mm in diam., black when mature. Fl. Mar-Apr, fr. May-Jul.

Distributed in Hunan (Yizhang, Tongdao), Guangdong, Guangxi and Southern Guizhou. Grows in broad-leaved forest in valleys and in limestone mountain.

常绿乔木。幼枝、芽鳞及嫩叶中脉均被黄棕色茸毛。叶革质，长披针形，先端渐尖，基部楔形，长12～18cm，上面深绿色无毛，下面被短毛，沿脉毛更密，中脉上面下凹，侧脉8～12对，小脉网状，常两面明显；叶柄长1～2cm，被茸毛，后渐脱落。圆锥花序数个丛生枝端，比叶短；花梗长约5mm；花淡黄色；花被片6，能育雄蕊9，花药4室。果实球形，无毛，径7～10mm，宿存花被片外曲，熟时紫黑色。花期3～4月，果期5～8月。

产新晃、通道、道县、江华、江永。生山谷沟边林中。分布于福建、广东、广西、贵州南部。性喜水湿，宜作护岸防堤树种。

Evergreen trees. Young twigs, bud scales and midribs of young leaves yellow-brown tomentose. Leaves leathery, long lanceolate, apex acuminate, base cuneate, 12-18cm long, dark green, glabrous adaxially, pubescent abaxially, especially along midrib and veins, midvein impressed adaxially, lateral veins 8-12 pairs, small veins reticulate, usually conspicuous on both surfaces; petioles 1-2cm long, tomentose, later gradual glabrous. Several panicles clustered at the top of branches, shorter than leaves; pedicels ca.5mm long; flowers pale yellow; tepals 6, fertile stamens 9, anthers 4-loculed. Fruit globose, glabrous, 7-10mm in diam., persistent tepals reflexed, violet-black when mature. Fl. Mar-Apr, fr. May-Aug.

Distributed in Hunan (Xinhuang, Tongdao, Daoxian, Jianghua, Jiangyong), Fujian, Guangdong, Guangxi and South Guizhou. Grows in forests in valleys and ravines. It likes water-wet environment, suitable as the bank protection and breakwater tree species.

绒毛润楠 *Machilus velutina* Champ. ex Benth.

常绿乔木。小枝、芽、叶下面和花序均密被锈色密茸毛。叶革质，矩圆形、卵状矩圆形或倒卵形，长5～13cm，先端短渐尖，基部楔形，侧脉8～11对，中脉、侧脉在上面略下凹，网脉不明显；叶柄长1～3cm。圆锥花序短，长2～3cm，密集生小枝顶端；花黄绿色；花被片6，宿存，反曲，有锈色茸毛，能育雄蕊9。果实球形，熟时黑色；果梗红色。花期10～11月，果期翌年2～3月。

产芷江、新晃、通道、城步、江华、永兴。生海拔500m以下山地林下、林缘，很少成林，常与其他树种混生。分布于浙江、江西、福建、广东、广西、贵州。中南半岛亦产。树皮及枝皮含胶质，可作制香原料；树形美观，新叶从银白变淡红，为优良园林观赏树种。

Evergreen trees. Branchlets, buds, lower surface of leaves and inflorescences densely rusty densely villous. Leaves leathery, oblong, ovate-oblong or obovate, 5-13cm long, apex short acuminate, base cuneate, lateral veins 8-11 pairs, midrib and lateral veins slightly impressed adaxially, reticulate veins inconspicuous; petioles 1-3cm long. Panicle short, 2-3cm long, crowded at the top of branchlets; flowers yellow-green; tepals 6, persistent, reflexed, rusty tomentose, fertile stamens 9. Fruit globose, black when mature; fruiting pedicels red. Fl. Oct-Nov, fr. Feb-Mar of 2nd year.

Distributed in Hunan(Zhijiang, Xinhuang, Tongdao, Chengbu, Jianghua, Yongxing), Zhejiang, Jiangxi, Fujian, Guangdong, Guangxi and Guizhou, also in Indo-China Peninsula. Grows in forests or forest margins in mountainous region, below alt. 500m. Rarely forms forests, usually mixed with other trees. Bark and branch skin contain gum, can be used as incense raw materials; it is an excellent ornamental tree species for its beautiful tree-shape and new leaves turning the light red from the silver-white color.

常绿乔木。小枝黄绿色，无毛。叶革质，椭圆形或卵状椭圆形，长8~16cm，先端渐尖，基部楔形，上面深绿色，下面绿白色，无毛，离基3出脉；叶柄长2.5~3.5cm。伞形花序腋生或枝侧生，总花梗极短或无；苞片多数，宽卵形，外面被柔毛；雄花花被片4，能育雄蕊6，花药4室；雌花花被片4，退化雄蕊6。果椭圆形，长约10mm，径约8mm；果梗稍增粗，长约7mm。花期9~10月，果期12月。

产新宁、通道、绥宁、道县、江永、江华、宜章、桂东、汝城。生低山林中。分布于福建、广东、广西、江西、云南东南部。叶大近轮生，可栽为观叶植物。

Evergreen trees. Twigs yellowish-green, glabrous. Leaves leathery, elliptic or ovate-elliptic, 8-16cm long, apex acuminate, base cuneate, dark green adaxially, green-white abaxially, glabrous, triplinerved; petioles 2.5-3.5cm long. Umbels axillary or lateral, peduncle very short or absent; bracts numerous, broadly ovate, outside puberulent; male flowers: tepals 4, fertile stamens 6, anthers 4-loculed; pistillate flowers: tepals 4, staminodes 6. Fruit elliptic, ca.10mm long, ca.8mm in diam.; fruiting pedicels slightly thickened, ca.7mm long. Fl. Sep-Oct, fr. Dec.

Distributed in Hunan(Xinning, Tongdao, Suining, Daoxian, Jiangyong, Jianghua, Yizhang, Guidong, Rucheng), Fujian, Guangdong, Guangxi, Jiangxi and Southeast Yunnan. Grows in low mountain forests. Leaf is large and subverticillate, and can be used as potted plants for foliage plants.

短蕊青藤　　*Illigera brevistaminata* Y. R. Li

木质藤本。叶互生，具3小叶；小叶长椭圆形或长卵圆形，近革质，先端尾状渐尖，基部圆形至亚圆形，两侧偏斜。聚伞花序腋生；花绿色，萼片5，长圆形；雄蕊5，长约2mm，粗壮；花柱与花丝近等长，约1mm，柱头波状扩大。果具4翅。花期9月。

产江华、江永、宜章。生低海拔石灰岩山地。分布于贵州东南部、广西北部、广东。

Woody vines. Leaves alternate, 3-foliolate; leaflets oblong or long ovoid, subleathery, apex caudate-acuminate, base rounded to subrounded, oblique on both sides. Cymes axillary; flowers green, sepals 5, oblong; stamens 5, ca.2mm long, stout; style ca. as long as filaments, ca.1mm, stigma corrugated expansion. Fruit 4-winged. Fl. Sep.

Distributed in Hunan(Jianghua, Jiangyong, Yizhang), Southeast Guizhou, North Guangxi and Guangdong. Grows in limestone mountain at low altitude.

樟叶木防己　*Cocculus laurifolius* DC.

常绿灌木。小枝具条纹，无毛。叶薄革质，椭圆状长圆形或长圆状披针形，长4～15cm，先端渐尖，基部渐狭，两面无毛，光亮，干时边缘呈微波状，基出脉3；叶柄长5～10mm。花单性，雌雄异株；聚伞花序或聚伞圆锥花序，腋生，长1～5cm；雄花萼片6；花瓣6，倒心形，顶端2裂；雄蕊6；雌花萼片和花瓣与雄花的相似；退化雄蕊6，微小；心皮3，无毛。核果近圆形，长6～7mm。花期春、夏，果期秋季。

产道县、东安、江华、江永、新宁。生石灰岩低山灌丛。分布于福建、台湾、广东、广西、贵州南部、云南。越南、老挝、缅甸、印度、日本亦产。叶片光鲜油亮，四季常绿，树姿优美雅致，可作园林绿化树种。

Evergreen shrubs. Twigs striate, glabrous. Leaves thinly leathery, elliptic-oblong or oblong-lanceolate, 4-15cm long, apex acuminate, base attenuate, glabrous on both surfaces, glossy, margin sinuolate when dry, basal veins 3; petioles 5-10mm long. Flowers unisexual, dioecious; cymes or cymose racemes, axillary, 1-5cm long; male flowers: sepals 6; petals 6, obcordate, apex 2-lobed; stamens 6; female flowers: sepals and petals similar to male flowers; staminodes 6, small; carpels 3, glabrous. Drupe subround, 6-7mm long. Fl. Spring and Summer, fr. Autumn.

Distributed in Hunan (Daoxian, Dongan, Jianghua, Jiangyong, Xinning), Fujian, Taiwan, Guangdong, Guangxi, South Guizhou and Yunnan, also in Vietnam, Laos, Burma, India and Japan. Grows in low altitude mountain thickets in limestone areas. It is suitable as the ornamental species because of its light and evergreen leaves, and elegant tree-shape.

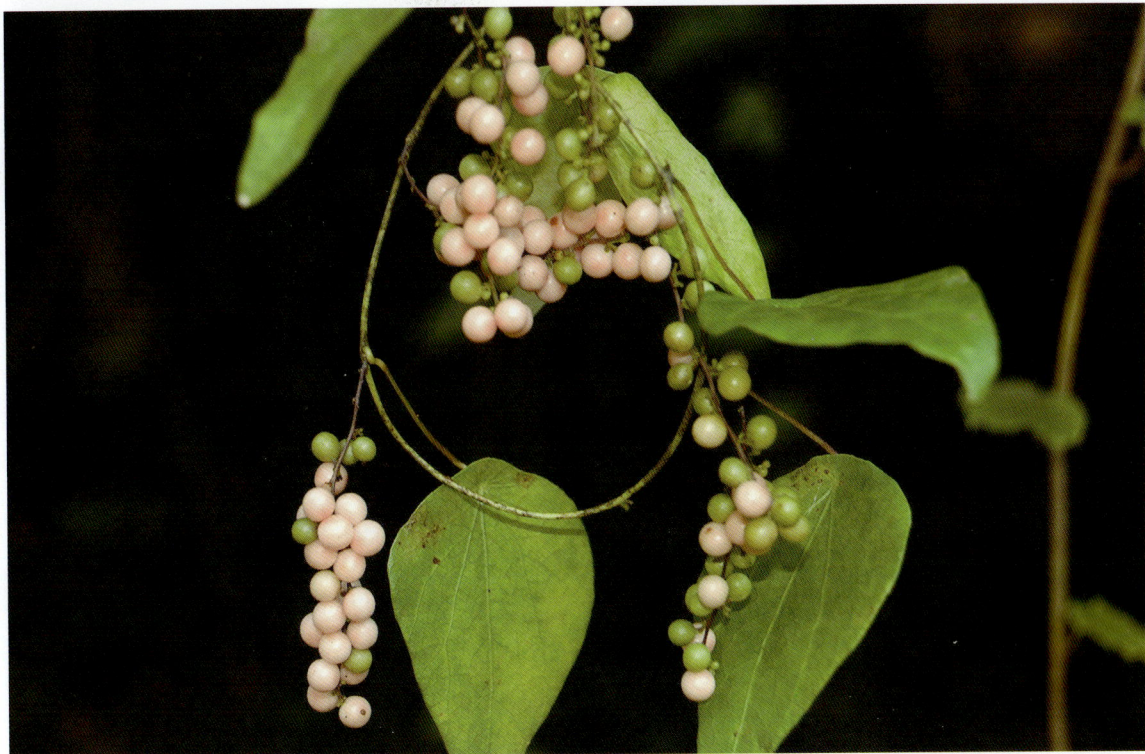

粉叶轮环藤　　*Cyclea hypoglauca* (Schauer) Diels

　　落叶木质藤本。叶片常明显盾状，阔卵状三角形至卵形，长2.5～7cm，先端渐尖，基部截平至圆，全缘而稍反卷，两面无毛或下面被稀疏长毛，掌状脉5～7；叶柄纤细，长1.5～4cm。花序腋生，雄花序为间断的穗状；雄花萼片4或5，分离；花瓣4～5，常合生成杯状，聚药雄蕊长1～1.2mm；雌花序较粗壮，呈总状，花序轴明显曲折；雌花萼片2，花瓣2，不等大。核果扁平，熟时红色，长约3.5mm。

　　产道县、东安、江华、江永、蓝山、宁远、汝城、新宁、宜章、资兴。生山坡疏林。分布于江西南部、福建、云南、广西、广东、海南。越南亦产。

Deciduous woody vines. Leaf blade usually conspicuously peltate, broadly ovate-triangular to ovate, 2.5-7cm long, apex acuminate, base truncate to rounded, margin entire and slightly reflexed, glabrous on both surfaces or sparsely hairy abaxially, palmately 5-7-veined; petioles slender, 1.5-4cm long. Inflorescences axillary, male inflorescences interrupted spike; male flowers: sepals 4or 5, free; petals 4-5, usually connate, cupular, synandria 1-1.2mm long; female inflorescence stout and racemose, inflorescence rachis apparent tortuosity; female flowers: sepals 2, petals 2, unequal in length. Drupe compressed, red when mature, ca.3.5mm long.

　　Distributed in Hunan (Daoxian, Dongan, Jianghua, Jiangyong, Lanshan, Ningyuan, Rucheng, Xinning, Yizhang, Zixing), South Jiangxi, Fujian, Yunnan, Guangxi, Guangdong and Hainan, also in Vietnam. Grows in sparse forests on mountain slopes.

黄花倒水莲 *Polygala fallax* Hemsl.

落叶灌木。叶膜质至纸质，椭圆状披针形至长圆形，长8~20cm，先端渐尖，基部楔形至钝形，全缘，两面无毛或疏生短柔毛，中脉上面凹陷，侧脉8~9对；叶柄长9~14mm，被短柔毛。总状花序顶生或腋生，长达30cm；萼片5，早落，具缘毛；花瓣3，黄色，龙骨瓣盔状，顶部具鸡冠状附属物，流苏状；雄蕊8。蒴果宽倒心形，绿黄色，径10~14mm。种子圆形，密被白色短柔毛。花期5~8月，果期8~10月。

产新宁、城步、道县、江华、江永、炎陵、桂东、资兴、蓝山、汝城、宜章。生低海拔山地林下。分布于江西、福建、广东、广西、香港、云南。

Deciduous shrubs. Leaves membranous to papery, elliptic-lanceolate to oblong, 8-20cm long, apex acuminate, base cuneate to obtuse, margin entire, glabrous or sparsely pubescent on both surfaces, midrib adaxially impressed, lateral veins 8-9 pairs; petioles 9-14mm long, pubescent. Racemes terminal or axillary, up to 30cm long; sepals 5, caducous, ciliate; petals 3, yellow, keel helmet-shaped, apically with cristate appendages, fimbriate; stamens 8. Capsule broadly obcordate, yellow green, 10-14mm in diam.. Seeds orbicular, densely white pubescent. Fl. May-Aug, fr. Aug-Oct.

Distributed in Hunan (Xinning, Chengbu, Daoxian, Jianghua, Jiangyong, Yanling, Guidong, Zixing, Lanshan, Rucheng, Yizhang), Jiangxi, Fujian, Guangdong, Guangxi, Hongkong and Yunnan. Grows in forests in mountain land at low altitude.

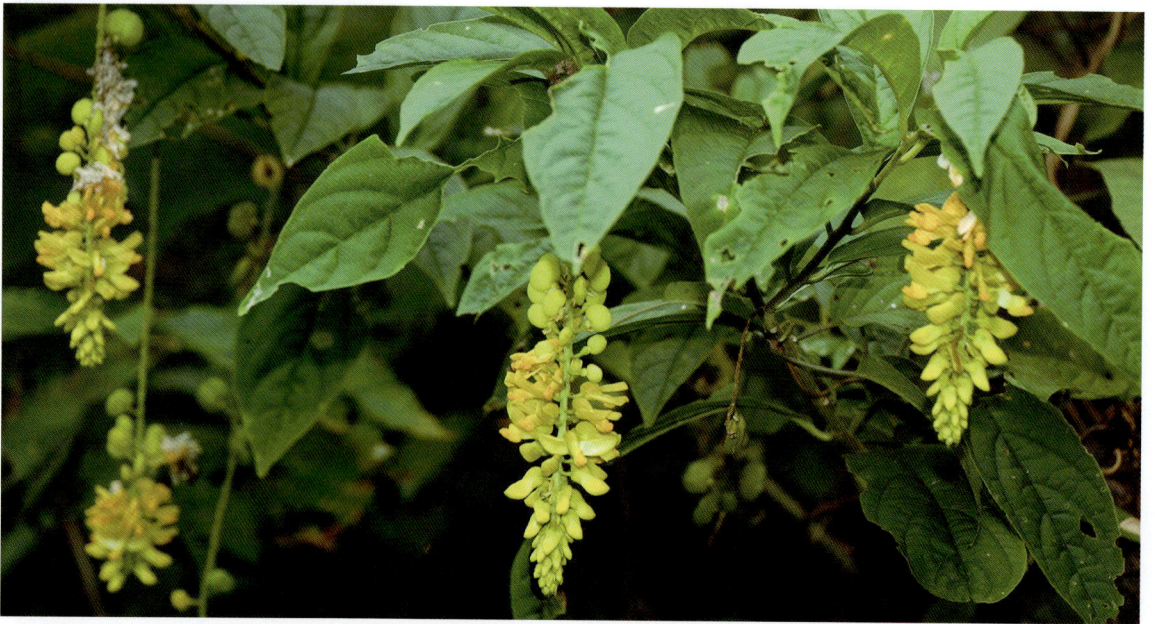

荷莲豆草 *Drymaria cordata* (L.) Willd. ex Schult.

一年生草本。茎蔓生，基部分枝，通常在下部节上生不定根。叶卵状心形，长1～1.5cm，先端凸尖，基出3～5脉；叶柄短；托叶小，白色，刚毛状。顶生聚伞花序；花梗细弱，短于花萼，被白色腺毛；萼片5，披针状卵形，边缘膜质，具3脉；花瓣5，白色，长约2.5mm，先端2深裂；雄蕊2～3(5)，短于萼片；花柱3，基部合生。蒴果卵形，长约2.5mm，3瓣裂；种子近圆形，密被瘤状突起。花期4～10月，果期6～12月。

产江华、江永。生低山溪边、山地草丛。分布于福建、台湾、广东、广西、海南、四川、贵州、云南。日本、印度、斯里兰卡、阿富汗、非洲南部亦产。

Herbs annual. Stem sprawling, branched basally, usually rooting at lower nodes. Leaves ovate-cordate, 1-1.5cm long, apex convex, basal veins 3-5; petioles short; stipules small, white, just hair. Terminal cymes; peduncle slender, shorter than calyx, white hairs; sepals 5, lanceolate-ovate, margin membranous, 3-veined; petals 5, white, ca.2.5mm long, deeply 2-cleft; stamens 2-3(5), shorter than sepals; ovary ovoid, styles 3, connate at base. Capsule ovoid, ca.2.5mm long, 3-valved; seeds suborbicular, densely tuberculate. Fl. Apr-Oct, fr. Jun-Dec.

Distributed in Hunan(Jianghua, Jiangyong), Fujian, Taiwan, Guangdong, Guangxi, Hainan, Sichuan, Guizhou and Yunnan, also in Japan, India, Sri Lanka, Afghanistan, South Africa. Grows in streamsides in low mountain, and in mountain grass.

常绿乔木。小枝顶端扁平。叶纸质至薄革质，无毛，椭圆形或长圆形，长4～16cm，先端锐尖或圆而微凹缺，基部宽楔形，全缘，上面中脉下凹，侧脉5～12对；叶柄长1～3cm。聚伞花序腋生；花小，白色；萼片5，卵状长椭圆形；花瓣5，宿存，宽圆形；花盘杯状；雄蕊10，伸出花冠外，长达2cm；子房5室，花柱长，柱头头状。蒴果卵状椭圆形，长2.5～3cm，果皮革质，熟时黑褐色，5瓣裂开；种子长圆形，具翅。花期5～6月，果期6～10月。

产通道、江华、江永、道县、汝城。生低山阔叶林中。分布于广西、广东、海南、云南、贵州。越南亦产。天然更新不良，系我国珍稀濒危植物。

Evergreen trees. Branchlets top compressed. Leaves papery to thinly leathery, glabrous, elliptic or oblong, 4-16cm long, apex acute or rounded and slightly emarginate, base broadly cuneate, margin entire, midrib impressed adaxially, lateral veins 5-12 pairs, petioles 1-3cm long. Cymes axillary; flowers small, white; sepals 5, ovate oblong; petals 5, persistent, broadly rounded; disk cupular; stamens 10, extend corollaceous outside, up to 2cm long; ovary 5-loculed, style long, stigmas capitate. Capsule ovate-elliptic, 2.5-3cm long, pericarp leathery, dark brown when mature, 5-valved; seeds oblong, winged. Fl. May-Jun, fr. Jun-Oct.

Distributed in Hunan(Tongdao, Jianghua, Jiangyong, Daoxian, Rucheng), Guangxi, Guangdong, Hainan, Yunnan and Guizhou, also in Vietnam. Grows in broad-leaved forests in low mountains. It is a rare and endangered plant in china for bad natural regeneration.

广东山龙眼 *Helicia kwangtungensis* W. T. Wang

常绿乔木。小枝和叶幼时被锈色短毛，后脱落。叶纸质或革质，长圆形或倒卵形，长10～26cm，先端短急尖、短渐尖或钝，基部楔形，边缘上半部疏生锯齿，有时全缘，侧脉及网脉不明显或叶背明显；叶柄长1～2.5cm。总状花序腋生，花序轴和花梗密被褐色短毛；花两性，淡黄色，花梗常双生，花被片4，花柱细长。果近球形，顶端具短尖，紫黑色，径1.5～2.5cm。花期6～7月，果期10～12月。

产宜章。生山地湿润常绿阔叶林中。分布于广西东南部、广东、江西南部、福建。

Evergreen trees. Branchlets and leaves rusty pubescent when young, glabrescent. Leaves papery or leathery, oblong to obovate, 10-26cm long, apex acute, shortly acuminate or obtuse, base cuneate, margin remotely serrulate on apical half, sometimes entire, and with reticulate veins inconspicuous or conspicuous abaxially; petioles 1-2.5cm long. Racemes axillary, inflorescence rachis and pedicels densely brown pubescent; flowers bisexual, pale yellow, pedicels often twins, tepals 4, style slender. Fruit subglobose, apex very shortly apiculate, purple-black, 1.5-2.5cm diameter. Fl. Jun-Jul, fr. Oct-Dec.

Distributed in Hunan(Yizhang), Southeast Guangxi, Guangdong, South Jiangxi and Fujian. Grows in humid evergreen broad-leaved forests in mountain land.

网脉山龙眼 *Helicia reticulata* W. T. Wang

常绿小乔木或灌木。叶革质或近革质，长圆形至倒卵状长圆形，长10～27cm，先端短渐尖或钝，基部楔形，两面无毛，边缘具疏生锯齿或细齿，侧脉8～10对，中脉、侧脉及网脉在两面均凸起或明显；叶柄长1～3cm。总状花序腋生，长10～14cm；花梗常双生，长3～5mm；花被管长1.3～1.6cm，白色或淡黄色，花盘4裂。果椭圆球形，径约1.5cm，顶端具短尖，果皮革质，黑色。花期5～7月，果期10～12月。

产新宁、通道、道县、江华、江永。生山地林缘、灌丛、疏林下。分布于云南东南部、贵州、广西、广东、江西、福建南部。种子含淀粉，有毒，水煮去毒后，可食用。

Evergreen small trees or shrubs. Leaves leathery or subleathery, oblong to obovate-oblong, 10-27cm long, apex shortly acuminate or obtuse, base cuneate, glabrous on both surfaces, margin sparsely serrate or serrulate, lateral veins 8-10 pairs, midrib, and with reticulate veins raised or conspicuous on both surfaces; petioles 1-3cm long. Racemes axillary, 10-14cm long; usually pedicels twins, 3-5mm long; perianth tube 1.3-1.6cm long, white or light yellow, disk 4-lobed. Fruit elliptic, ca.1.5cm in diam., apex mucronate, pericarp leathery, black. Fl. May-Jul, fr. Oct-Dec.

Distributed in Hunan(Xinning, Tongdao, Daoxian, Jianghua, Jiangyong), Southeast Yunnan, Guizhou, Guangxi, Guangdong, Jiangxi and South Fujian. Grows in forest margins, thickets and sparse forests in mountain land. Seeds contain starch, toxic, edible after boiled detoxificatipairs.

短柄山桂花　*Bennettiodendron brevipes* Merr.

常绿灌木或小乔木。幼枝密被灰褐色短柔毛，后脱落近无毛。叶革质，长圆状倒披针形至倒卵状披针形，长5～12cm，先端短渐尖，基部楔形，两面无毛，侧脉6～12对；叶柄粗壮，长4～12mm，密被黄色柔毛。圆锥花序顶生，长4～9cm，被黄色柔毛。萼片卵形，边缘具睫毛；花瓣不存在；雄花具多数雄蕊；雌花具多数退化雄蕊；花柱3，线形。浆果球形，径3～4mm，熟时朱红色。花期春季，果期7～10月。

产通道、江华、江永、宜章、资兴。生山坡阔叶林下或石灰岩林下。分布于江西南部、广东、广西、贵州、云南、海南。

Evergreen shrubs or small trees. Young branchlets densely gray brown pubescent, later gradual glabrous. Leaves leathery, oblong-oblanceolate to obovate-lanceolate, 5-12cm long, apex shortly acuminate, base cuneate, glabrous on both surfaces, lateral veins 6-12 pairs; petioles stout, 4-12mm long, densely yellow villous. Panicles terminal, 4-9cm long, yellow pubescent. Sepals ovate, margin ciliate; petals absent; male flowers with numerous stamens; female flowers with numerous staminodes; styles 3, linear. Berry globose, 3-4mm in diam., scarlet when mature. Fl. Spring, fr. Jul-Oct.

Distributed in Hunan(Tongdao, Jianghua, Jiangyong, Yizhang, Zixing), South Jiangxi, Guangdong, Guangxi, Guizhou, Yunnan and Hainan. Grows in broad-leaved forests on mountain slopes or in limestone forests.

产新宁、江永、宁远、宜章。生石灰岩山地或低山丘陵。分布于广东、广西、海南、贵州、云南、福建南部、江西南部。印度及中南半岛亦产。

Evergreen shrubs or small trees. Branchlets pubescent. Leaves leathery, oblong to long elliptic-lanceolate, 6-14cm long, apex acuminate, base cuneate to broadly cuneate, grayish-yellow pubescent abaxially, margin coarsely serrate, lateral veins 5-8 pairs; Petioles 6-8mm long. Unisexual; racemes axillary, or sometimes branched, many flowered; pedicels 2-4mm long, sepals 4, apetalous; male flowers with numerous stamens, disk 4-8-valved, located in stamens peripheral; female flowers with annular disk, styles short, stigma obscurely 2–cleft. Berry globose, 3-6mm in diam., style persistent, brown when mature.

Distribution in Hunan(Xinning, Jiangyong, Ningyuan, Yizhang), Guangdong, Guangxi, Hainan, Guizhou, Yunnan, South Fujian and South Jiangxi, also in India and Indo-China Peninsula. Grows in low limestone mountain forests or low mountain and hill forests.

常绿灌木或小乔木。小枝被短柔毛。叶革质，长圆形至长椭圆状披针形，长6～14cm，先端渐尖，基部楔形至宽楔形，下面被灰黄色短柔毛，边缘具粗锯齿，侧脉5～8对；叶柄长6～8mm。雌雄异株；总状花序腋生，或有分枝，有多数花；花梗长2～4mm，萼片4，无花瓣；雄花具多数雄蕊，花盘4～8裂，位于雄蕊外围；雌花具环状花盘，花柱短，柱头不明显2裂。浆果球形，径3～6mm，花柱宿存，熟时褐色。

天料木　　*Homalium cochinchinense* (Lour.) Druce

落叶灌木或小乔木。小枝幼时密被黄色短柔毛，老时毛脱落。叶纸质，椭圆形至倒卵状矩圆形，长6～13cm，先端锐尖或短渐尖，基部楔形或宽楔形，边缘具疏锯齿，两面沿脉被短柔毛，中脉上面下凹，侧脉7～9对，弯拱相连；叶柄粗短，长2～3mm，被黄色短柔毛。总状花序穗状，腋生，长8～15cm；花白色，萼筒陀螺形，长2～2.5mm，疏生长柔毛，裂片6～8，线形到狭披针形；花瓣狭倒披针状线形到狭匙形。蒴果倒圆锥状，近无毛。花期4～5月，果期9～12月。

产新宁、道县、江华、江永、宜章。生山地次生阔叶林中。分布于江西、福建、台湾、广东、广西、海南。越南亦产。木材坚重，耐腐，为广东、广西、海南的名贵用材树种。

Deciduous shrubs or small trees. Branchlets densely yellow pubescent when young, glabrescent when old. Leaves papery, elliptic to obovate-oblong, 6-13cm long, apex acute or short acuminate, base cuneate or broadly cuneate, margin sparsely serrate, along veins pubescent on both surfaces, midvein impressed adaxially, lateral veins 7-9 pairs, curved arch connected; petioles short, 2-3mm long, yellow pubescent. Racemes spicate, axillary, 8-15cm long; flowers white, calyx tube turbinate, 2-2.5mm long, sparsely villous, lobes 6-8, linear to narrowly oblanceolate; petals narrowly oblanceolate-linear to narrowly spatulate. Capsule obconic, subglabrous. Fl. Apr-May, fr. Sep-Dec.

Distributed in Hunan(Xinning, Daoxian, Jianghua, Jiangyong, Yizhang), Jiangxi, Fujian, Taiwan, Guangdong, Guangxi and Hainan, also in Vietnam. Grows in secondary broad-leaved forests in mountain land. It is a precious timber tree in Guangdong, Guangxi and Hainan, for the timber is tough and thick and corrosion resistant.

广东西番莲 · *Passiflora kwangtungensis* Merr.

多年生草质藤本。茎纤细，光滑。叶膜质，披针形至长圆状披针形，长6~13cm，先端长渐尖，基部心形，全缘，下面无腺体，基出3脉，侧脉弯拱上升，网脉疏散而不显著；叶柄中部或上部具2枚小盘状腺体。花序无梗，成对生于纤细卷须的两侧，有1~2花；花小，白色；萼片5，外侧先端不具角状附属器；花瓣5，与萼片相似，副花冠裂片1轮，丝状；内花冠褶状。浆果球形，径1~1.5cm，无毛。花期3~5月，果期6~7月。

产江永、靖州、汝城、新宁、双牌。生山坡林边、灌丛中。分布于广东北部、广西东北部、江西东南部。

Perennial herbaceous vines. Stem slender, smooth. Leaves membranous, lanceolate to oblong-lanceolate, 6-13cm long, apex long acuminate, base cordate, margin entire, without glands abaxially, basal veins 3, lateral veins curved arch rise, reticulate veins evacuation and not significant; petioles middle or upper with 2 small disciform glands. Inflorescences sessile, paired in slender tendrils on both sides, with 1-2 flowers; flowers small, white; sepals 5, lateral apex without hornlike appendages; petals 5, similar to sepals, deputy corolla lobes 1 round, filiform; inner corolla pleated. Berry globose, 1-1.5cm in diam., glabrous. Fl. Mar-May, fr. Jun-Jul.

Distributed in Hunan(Jiangyong, Jingzhou, Rucheng, Xinning, Shuangpai), North Guangdong, Northeast Guangxi and Southeast Jiangxi. Grows in forest margins and thickets on mountain slopes.

心叶毛蕊茶　*Camellia cordifolia* (Metc.) Nakai

常绿灌木。嫩枝密生黄褐色或淡黄色长柔毛。叶革质，矩圆状椭圆形或披针形，长5～10cm，先端渐尖或尾状渐尖，基部圆形至微心形，边缘具细锯齿，下面被疏柔毛，中脉尤多；叶柄长2～4mm，密生长柔毛。花乳白色，单生叶腋；苞片4～5，宿存；萼片5，卵形，宿存；花瓣5，近圆形；花丝合生成雄蕊管，分离花丝被毛；子房和花柱密被毛，花柱先端3浅裂。蒴果近球形，径1～1.7cm。花期11月至翌年1月，果期9～10月。

产通道、城步、江华、江永、宁远、武冈、道县、宜章、资兴。生山地疏林下或林缘。分布于台湾、广东、广西、江西、福建、贵州、云南。越南北部亦产。

Evergreen shrubs. Young twigs densely yellow-brown or light yellow villous. Leaves leathery, oblong-elliptic or lanceolate, 5-10cm long, apex acuminate or caudate-acuminate, base rounded or slightly cordate, margin serrulate, sparsely puberulent abaxially, especially midrib more; petioles 2-4mm long, densely villous. Flower creamy white, solitary and axillary; bracts 4-5, persistent; sepals 5, ovate, persistent; petals 5, suborbicular; filaments connate into staminal tube, free filaments hairy; ovary and style densely hairy, style apex 3-lobed. Capsule subglobose, 1-1.7cm in diam.. Fl. Nov-Jan of 2nd year, fr. Sep-Oct.

Distributed in Hunan(Tongdao, Chengbu, Jianghua, Jiangyong, Ningyuan, Wugang, Daoxian, Yizhang, Zixing), Taiwan, Guangdong, Guangxi, Jiangxi, Fujian, Guizhou and Yunnan, also in North Vietnam. Grows in sparse forests or forest margins in mountain land.

糙果茶　　　*Camellia furfuracea* (Merr.) Coh. Stusrt

常绿灌木。嫩枝无毛。叶革质或厚革质，长圆形至披针形，长8～15cm，先端渐尖，基部楔形或钝，两面无毛，侧脉7～8对，与网脉在上面下凹；叶柄长5～8mm，无毛。花1～2朵顶生及腋生，无花梗，白色；苞片及萼片7～9，薄膜质，背面略被毛；花瓣7～8，倒卵形；雄蕊长1.3～1.5cm，花丝管长5～6mm，无毛，子房被长柔毛，花柱3，分离，被毛。蒴果球形或扁球形，径2.5～4cm，表面多瘤点，3瓣裂开，无宿存苞片或萼片。花期9～12月，果实秋后成熟。

产通道、宜章。生海拔200～500m的山坡灌丛或林缘。分布于广东、广西、福建、江西。

Evergreen shrubs. Young twigs glabrous. Leaves leathery or thickly leathery, oblong to lanceolate, 8-15cm long, apex acuminate, base cuneate or obtuse, glabrous on both surfaces, lateral veins 7-8 pairs, and with reticulate veins impressed adaxially; petioles 5-8mm long, glabrous. 1-2 flowers terminal and axillary, sessile, white; bracts and sepals 7-9, membranous, slightly hairy abaxially; petals 7-8, obovate; stamens 1.3-1.5cm long, filament tube 5-6mm long, glabrous; ovary villous, styles 3, free, hairy. Capsule globose or oblate, 2.5-4cm in diam., much verruculose on surface, 3-valved, without persistent bracts and sepals. Fl. Sep-Dec, fr. Autumn.

Distributed in Hunan(Tongdao, Yizhang), Guangdong, Guangxi, Fujian and Jiangxi. Grows in thickets or forest margins on mountain slopes at alt. 200-500m.

柳叶毛蕊茶（柳叶山茶） *Camellia salicifolia* Champ.ex Benth.

常绿灌木。幼枝纤细，密生黄褐色长柔毛，后变无毛。叶近革质，披针形，长4～10cm，先端尾状渐尖，基部圆形，下面被长柔毛；叶柄长2～3mm，密生黄褐色长柔毛。花白色，顶生或腋生；花梗和萼密生长柔毛；苞片和萼片窄披针形，萼片宿存；雄蕊多数，外轮花丝下部合生约达2/3，密生长柔毛；子房、花柱密被毛，花柱先端3浅裂。蒴果近球形，径1.5～2.2cm。花期9～11月。

产汝城、宜章。生山地阔叶疏林中。分布于台湾、广东、广西、香港、江西、福建。

Evergreen shrubs. Young twigs slender, densely yellow-brown villous, later becoming glabrous. Leaves subleathery, lanceolate, 4-10cm long, apex caudate-acuminate, base rounded, villous abaxially; petioles 2-3mm long, densely yellow-brown villous. Flowers white, terminal or axillary; pedicels and calyx densely villous; bracts and sepals narrow lanceolate, sepals persistent; stamens numerous, outer filaments connate ca.2/3, densely villous; ovary and style densely hairy, styles 3, apex 3-lobed. Capsule subglobose, 1.5-2.2cm in diam.. Fl. Sep-Nov.

Distributed in Hunan(Rucheng, Yizhang), Taiwan, Guangdong, Guangxi, Hongkong, Jiangxi and Fujian. Grows in broad-leaved sparse forests in mountain land.

茶 梨　　*Anneslea fragrans* Wall.

常绿乔木。当年小枝红棕色，无毛。叶厚革质，簇生小枝顶端，披针形或长圆状披针形，长4.5~15cm，全缘或具疏锯齿，下面苍绿色并散生黑色腺点，侧脉不明显；叶柄粗壮，长2~3cm。花腋生，由数花至10多花排成伞房花序；花梗长3~6cm；萼片红色，长1~1.5cm，基部稍合生，边缘具腺；花瓣淡黄色，宽卵形，长约1.5cm；子房半下位，2~3室，花柱红色。浆果球形至椭圆形，径2~3.4cm，花萼宿存且收缩。花期12月至翌年3月，果期7~11月。

产通道、洞口、新宁、城步、江永、宁远、道县、宜章。生山坡林下。分布于福建、江西、广东、广西、贵州、云南、台湾、海南。柬埔寨、老挝、马来西亚、缅甸、泰国、越南亦产。可作园景树。

Evergreen trees. Current year branchlets reddish brown, glabrous. Leaves thick leathery, clustered at the top of branchlets, lanceolate or oblong-lanceolate, 4.5-15cm long, margin entire or with sparse serrated, pale green and scattered black glandular punctate abaxially, lateral veins inconspicuous; petioles stout, 2-3cm long. Flowers axillary, several to more than 10 in a corymb; pedicels 3-6cm long; sepals red, 1-1.5cm long, base slightly connate, margin glands; petals pale yellow, broadly ovate, ca.1.5cm long; ovary half inferior, 2-3-loculed, style red. Berry globose to elliptic, 2-3.4cm in diam., calyx persistent and contraction. Fl. Dec-Mar of 2nd year, fr. Jul-Nov.

Distributed in Hunan(Tongdao, Dongkou, Xinning, Chengbu, Jiangyong, Ningyuan, Daoxian, Yizhang), Fujian, Jiangxi, Guangdong, Guangxi, Guizhou, Yunnan, Taiwan and Hainan, also in Cambodia, Laos, Malaysia, Myanmar, Thailand and Vietnam. Grows in forests on mountain slopes. It can be used as landscape trees.

尖萼毛枴　　*Eurya acutisepala* Hu et L. K. Ling

常绿灌木。嫩枝圆柱形，黄褐色，密被短柔毛；多年生小枝灰褐色，无毛；顶芽披针形，密被黄褐色丝状毛。叶薄革质，长圆形或倒披针状长圆形，长5~8cm，先端长渐尖，尾长1~1.5cm，基部阔楔形或楔形，边缘密生细锯齿；叶柄长2~3.5mm，被短柔毛。2~3花腋生，花梗长1.5~2.5mm，疏被短柔毛；萼片卵形，先端尖，无毛；雄花较大，雄蕊约15；雌花较小，子房卵形，密被柔毛，花柱顶端3裂。果实卵状椭圆形，熟时紫黑色，长约3.5~4.5mm，疏被柔毛。花期10~11月，果期翌年6~8月。

产城步、江华、宁远、新晃、通道、新宁、道县、宜章、资兴。生山坡林下。分布于浙江南部、江西、福建、广东、广西、贵州、云南。

Evergreen shrubs. Young twigs terete, yellow-brown, densely pubescent; perennial branchlets gray brown, glabrous; terminal buds lanceolate, densely yellowish-brown filiform trichomes. Leaves thinly leathery, oblong or oblanceolate-oblong, 5-8cm long, apex long acuminate, tail 1-1.5cm long, base broadly cuneate or cuneate, margin densely serrulate; petioles 2-3.5mm long, pubescent. 2-3 flowers axillary, pedicels 1.5-2.5mm long, sparsely pubescent; sepals ovate, apex acute, glabrous; male flowers: larger, stamens ca.15; female flowers: smaller, ovary ovoid, densely pilose, style apically 3-lobed. Fruit ovate-elliptic, purple-black when mature, ca.3.5-4.5mm long, sparsely pilose. Fl. Oct-Nov, fr. Jun-Aug of 2nd year.

Distributed in Hunan(Chengbu, Jianghua, Ningyuan, Xinhuang, Tongdao, Xinning, Daoxian, Yizhang, Zixing), South Zhejiang, Jiangxi, Fujian, Guangdong, Guangxi, Guizhou and Yunnan. Grows in forests on mountain slopes.

米碎花 *Eurya chinensis* R. Br.

常绿小灌木。嫩枝具2棱，与顶芽被短柔毛。叶薄革质，倒卵形或倒卵状椭圆形，长2~5.5cm，先端钝，边缘密生细锯齿，侧脉6~8对；叶柄长2~3mm。花白色至黄绿色，1~4朵腋生；萼片卵形，无毛；雄花苞片细小，花瓣倒卵形，长3~3.5mm，雄蕊15；雌花花瓣卵形，长2~2.5mm，子房无毛，花柱长1.5~2mm，顶端3浅裂。果实圆球形，径约4mm，熟时黑色。花期11~12月，果期翌年6~7月。

产通道、江华、宜章。生低海拔荒山灌丛、草地、河边。分布于江西、福建、台湾、广东、广西。越南、缅甸、印度、斯里兰卡、印度尼西亚亦产。

Evergreen small shrubs. Young twigs 2-angled, and with terminal buds pubescent. Leaves thinly leathery, obovate or obovate-elliptic, 2-5.5cm long, apex obtuse, margin densely serrulate, lateral veins 6-8 pairs; petioles 2-3mm long. Flower white to yellowish-green, 1-4 flowers axillary; sepals ovate, glabrous; male flowers: bracts small, petals obovate, 3-3.5mm long, stamens 15; female flowers: petals ovate, 2-2.5mm long, ovary glabrous, style 1.5-2mm long, apex 3-lobed. Fruit globose, ca.4mm in diam., black when mature. Fl. Nov-Dec, fr. Jun-Jul of 2nd year.

Distributed in Hunan (Tongdao, Jianghua, Yizhang), Jiangxi, Fujian, Taiwan, Guangdong and Guangxi, also in Vietnam, Burma, India, Sri Lanka, Indonesia. Grows in barren mountain shrubs, grassland, riverside at low altitude.

二列叶柃　*Eurya distichophylla* Hemsl.

常绿灌木。嫩枝圆柱形，被黄褐色长丝毛。叶坚纸质，矩圆状披针形，长3～6cm，宽8～15mm，先端锐尖，基部圆形而略不对称，边缘生细锯齿，下面被长丝毛；叶柄极短。1～3花腋生，萼片卵形，被毛，花瓣长2～2.5mm，顶端白或淡绿色，基部深紫色；雄花雄蕊15；雌花子房被长丝毛，花柱3深裂。果实圆球形，径4～5mm，被黄褐色长丝毛。花期10～12月，果期翌年6～7月。

产宜章。生山地灌丛、林下、水边。分布于福建、广东、广西、江西。越南亦产。

Evergreen shrubs. Young twigs terete, yellow-brown filiform trichomes. Leaves hard papery, oblong-lanceolate, 3-6mm × 8-15mm, apex acute, base rounded and slightly asymmetric, margin serrulate, filiform trichomes abaxially; petioles very short. 1-3 flowers axillary, sepals ovate, hairy, petals 2-2.5mm long, apex white or pale green, base dark purple; male flowers: stamens 15; female flowers: ovary filaments hairy, style 3-parted. Fruit globose, 4-5mm in diam., yellow-brown filiform trichomes. Fl. Oct-Dec, fr. Jun-Jul of 2nd year.

Distributed in Hunan(Yizhang), Fujian, Guangdong, Guangxi and Jiangxi, also in Vietnam. Grows in shrubs, forests and watersides in mountain land.

常绿灌木。嫩枝圆柱形，密生黄褐色长丝毛。叶革质，互生，长圆状披针形，长5～12cm，先端渐尖，基部圆形或微心形，下面被黄褐色长丝毛，边缘具细锯齿，侧脉上面不明显；叶柄极短，长1～2mm，密被黄色毛。1～3花腋生；花梗长1mm，被毛；雄花萼片卵形，先端圆，外面被长丝毛；花瓣卵形，基部合生，雄蕊15～19枚；雌花子房卵形，被长丝毛。果实卵球形，径约5mm，被黄褐色长丝毛。花期10～12月，果期翌年6～7月。

产江华、道县。生山谷、林缘、路旁。分布于广西、广东，其中广西较多。

Evergreen shrubs. Young twigs terete, densely yellow-brown filiform trichomes. Leaves leathery, alternate, oblong-lanceolate, 5-12cm long, apex acuminate, base rounded or slightly cordate, yellow-brown filiform trichomes abaxially, margin serrulate, lateral veins inconspicuous adaxially; petioles very short, 1-2mm long, densely yellow hair. 1-3 flowers axillary; pedicels 1mm long, hairy; male flowers: sepals ovate, apex rounded, outside filiform trichomes; petals ovate, connate at base, stamens 15-19; female flowers: ovary ovoid, filiform trichomes. Fruit ovoid, ca.5mm in diam., yellow-brown filiform trichomes. Fl. Oct-Dec, fr. Jun-Jul of 2nd year.

Distributed in Hunan(Jianghua, Daoxian), Guangxi, Guangdong, and more common in Guangxi among them. Grows in valleys, forest margins and roadsides.

单耳柃 *Eurya weissiae* Chun

常绿灌木。嫩枝圆柱形，密被黄褐色披散长柔毛。叶革质，长圆形或椭圆状长圆形，长4～8cm，基部斜耳形抱茎，两侧耳片圆形，通常下侧较大，边缘密生细锯齿，中脉、侧脉及网脉在上面明显下凹；叶柄极短或近无柄。花1～3腋生，为一细小叶状总苞所包裹；花梗短，长约1mm，被柔毛。果实圆球形，径4～5mm，熟时蓝黑色。花期9～11月，果期11月至翌年1月。

产通道、城步、汝城、桂东、宜章、炎陵、永兴、资兴。生山坡林下。分布于浙江南部、江西南部、福建、广东、广西、贵州。

Evergreen shrubs. Young twigs terete, densely yellow-brown villous. Leaves leathery, oblong or elliptic-oblong, 4-8cm long, base obliquely auriculate and amplexicaul, auricles round on both sides, usually lower auricle larger, densely serrulate, midrib, lateral veins and reticular veins significantly impressed adaxially; petioles very short or subsessile. 1-3 flowers axillary, wrapped by a small leaflike involucre bract; pedicels short, ca.1mm long, pilose. Fruit globose, 4-5mm in diam., bluish black when mature. Fl. Sep-Nov, fr. Nov-Jan of 2nd year.

Distributed in Hunan(Tongdao, Chengbu, Rucheng, Guidong, Yizhang, Yanling, Yongxing, Zixing), South Zhejiang, South Jiangxi, Fujian, Guangdong, Guangxi and Guizhou. Grows in forests on mountain slopes.

常绿乔木。嫩枝及叶均无毛。叶革质，长圆形，长9～14cm，先端渐尖，基部楔形，上面干后发亮，下面无毛，侧脉9～12对，边缘具细锯齿；叶柄长6～12mm，无毛。花单生于枝顶叶腋，淡黄色，径4～5cm，花梗长5～10mm，被毛；萼片5，卵圆形，外面被灰褐色柔毛；花瓣5，倒卵形，外面被灰色绢毛；子房被毛，花柱长8mm，无毛。蒴果卵形或椭圆形，3～5室，径2～3.5cm，果皮干燥时1～2mm厚，外被褐色茸毛。花期6～7月，果期9～10月。

产通道、宜章、汝城。生山谷林中、山坡疏林。分布于江西南部、福建、广东、广西、贵州。

Evergreen trees. Young twigs and leaves glabrous. Leaves leathery, oblong, long 9-14cm, apex acuminate, base cuneate, after dry shiny adaxially, glabrous abaxially, lateral veins on 9-12, margin serrulate; petioles 6-12mm, glabrous. Flowers solitary and axillary at the top of branches, pale yellow, 4-5cm in diam., pedicels 5-10mm long, hairy; sepals 5, ovate-orbicular, outside gray brown pilose; petals 5, obovate, outside gray sericeous; ovary hairy, style 8mm long, glabrous. Capsule ovate or elliptic, 3-5-loculed, 2-3.5cm in diam., pericarp 1-2mm thick when dry, outside tan-colared hairs. Fl. Jun-Jul, fr. Sep-Oct.

Distributed in Hunan(Tongdao, Yizhang, Rucheng), South Jiangxi, Fujian, Guangdong, Guangxi and Guizhou. Grows in valley forests, and in sparse forests on mountain slopes.

五列木 *Pentaphylax euryoides* Gardn. et Champ.

常绿灌木或小乔木。小枝淡灰棕色，圆柱状，无毛。叶革质，卵形，长圆状卵形，或长圆状披针形，全缘，稍外卷，长5～9cm，两面无毛，先端尾状，基部圆形到宽楔形，侧脉不明显，叶柄1～1.5cm，上面具槽；花梗长约0.5mm；萼片5，圆形；花瓣长圆状披针形至倒披针形，长4～5mm，先端微凹，2半裂；雄蕊5；花柱先端5裂。蒴果紫褐色至黑褐色，椭圆形，长6～9mm，熟后5裂。花期6月，果期10月。

产道县、江华、江永、蓝山、宁远、宜章、汝城、炎陵、通道。生海拔500m以下石灰岩山丘。分布于福建南部、广东、广西、贵州南部、海南、江西南部、云南。印度尼西亚、马来西亚、越南亦产。

Evergreen shrubs or small trees. Branchlets gray brown, terete, glabrous. Leaves leathery, ovate, oblong-ovate, or oblong-lanceolate, margin entire and slightly revolute; 5-9cm long, glabrous on both surfaces, apex caudate, base rounded to broadly cuneate, lateral veins obscure, petioles 1-1.5cm long, adaxially grooved. pedicels ca.0.5mm long; sepals 5, orbicular; petals oblong-lanceolate to oblanceolate, 4-5mm long, glabrous, apex retuse to 2-cleft; stamens 5; style apically 5-lobed. Capsule purplish-brown to dark brown, elliptic, 6-9mm long, 5-valved after dry. Fl. Jun, fr. Oct.

Distributed in Hunan(Daoxian, Jianghua, Jiangyong, Lanshan, Ningyuan, Yizhang, Rucheng, Yanling, Tongdao), South Fujian, Guangdong, Guangxi, South Guizhou, Hainan, South Jiangxi and Yunnan, also in Indonesia, Malaysia, and Vietnam. Grows in limestone hills below alt. 500m.

桃金娘　*Rhodomyrtus tomentosa* (Ait.) Hassk

常绿小灌木。幼枝密被柔毛。叶革质，对生，椭圆形或倒卵形，长3～6cm，先端微凹，基部楔形，全缘，下面被灰白色短茸毛，离基3出脉，侧脉4～7对；叶柄4～7mm。聚伞花序腋生，有花1～3朵；花紫红色，径2～4cm；小苞片2，卵形；萼筒钟形，长5～6mm，裂片5，圆形；花瓣5，倒卵形，长约1.5cm；雄蕊多数；子房下位。浆果卵形，径1～1.4cm，熟时暗紫色。花期4～5月，果期7～9月。

产通道、江永、汝城、桂东。生丘陵红壤灌丛中。分布于福建、台湾、广东、广西、云南、贵州、江西。印度至菲律宾，日本南部亦产。果味甜可食。

Evergreen small shrubs. Young twigs densely pilose. Leaves leathery, opposite, elliptic or obovate, 3-6cm long, apex retuse, base cuneate, margin entire, grey velutinous abaxially, triplinerved, lateral veins 4-7 pairs; petioles 4-7mm long. Cymes axillary, with 1-3 flowers; flowers purplish-red, 2-4cm in diam.; bracteoles 2, ovate; calyx tube campanulate, 5-6mm long, lobes 5, oblong; petals 5, obovate, ca.1.5cm long; stamens numerous; ovary inferior. Berry ovoid, 1-1.4cm in diam., dark purple when mature. Fl. Apr-May, fr. Jul-Sep.

Distributed in Hunan(Tongdao, Jiangyong, Rucheng, Guidong), Fujian, Taiwan, Guangdong, Guangxi, Yunnan, Guizhou and Jiangxi, also in India to Philippines, South Japan. Grows in thickets in red soil hilly. Fruit is sweet and edible.

常绿灌木。幼枝常4棱形，纤细，被灰色柔毛。叶近革质，对生，椭圆状披针形或卵状椭圆形，长3～7cm，先端渐尖，具0.5～1cm的尖头，基部楔形至钝形，侧脉不甚明显；叶柄长3～5mm，无毛。聚伞花序腋生，长约2cm；花轴疏被绢毛；花白色，3数，萼筒钟形，被绢毛；花瓣倒卵形，边缘具缘毛；雄蕊多数，离生；子房下位。浆果球形，径4～5mm，熟时红色。花期3～5月。

产江永、通道。生低海拔石灰岩山地。分布于广东、广西、贵州、台湾、海南。越南亦产。

Evergreen shrubs. Branchlets often 4-angled, slender, gray pilose. Leaves subleathery, opposite, elliptic-lanceolate or ovate-elliptic, 3-7cm long, apex acuminate with a 0.5-1cm acumen, base cuneate to obtuse, lateral veins inconspicuous; petioles 3-5mm long. Cymes axillary, ca.2cm long; axes sparsely sericeous; flowers white, 3-merous, calyx tube campanulate, sericeous; petals obovate, margin ciliate; stamens numerous, free; ovary inferior. Berry globular, 4-5mm in diam., red when mature. Fl. Mar-May.

Distributed in Hunan(Jiangyong, Tongdao), Guangdong, Guangxi, Guizhou, Taiwan and Hainan, also in Vietnam. Grows in limestone mountain land at low altitude.

常绿乔木。嫩枝有4棱。叶革质，对生，椭圆形，长4～7cm，先端锐尖或稍钝，基部宽楔形，两面被腺体，腺体在下面突起，侧脉多而密，侧脉相隔1.5～2mm，在上面不明显，在距叶缘约1mm处连结成边脉；叶柄长3～5mm。聚伞花序顶生或近顶生，长1.5～2.5mm；花梗长2～5mm；花芽倒卵球形，长4mm；萼管倒圆锥形，长2.5～3mm，萼片4，短三角形；花瓣白色，分离，倒卵形，长2.5mm。果实球形，径6～7mm。花期6～8月。

产汝城、通道。生海拔200m低山林中、山谷。分布于湖北、四川、贵州、江西、浙江、福建、广东、广西、海南。

Evergreen trees. Young twigs 4-angled. Leaves leathery, opposite, elliptic, 4-7cm long, apex acute or slightly obtuse, base broadly cuneate, with glands on both surfaces, glands raised abaxially, lateral veins numerous and dense, 1.5-2mm apart, inconspicuous adaxially, intramarginal veins ca.1 mm from margin; petioles 3-5 mm long. Cymes terminal or subterminal, 1.5-2.5mm long; pedicels 2-5mm long; flower buds obovoid, 4mm long; calyx tube obconic, 2.5-3mm long, sepals 4, short triangular; petals white, free, obovate, 2.5mm long. Fruit globose, 6-7mm in diam.. Fl. Jun-Aug.

Distributed in Hunan(Rucheng, Tongdao), Hubei, Sichuan, Guizhou, Jiangxi, Zhejiang, Fujian, Guangdong, Guangxi and Hainan. Grows in low mountain forests and valleys at alt. 200m.

轮叶蒲桃　　*Syzygium grijsii* (Hance) Merr. et Perry

常绿灌木。嫩枝纤细，具4棱，干后黑褐色。叶革质，细小，常3叶轮生，狭窄长圆形或狭披针形，长1.5～2cm，先端钝或略尖，基部楔形，上面干后暗褐色，下面具多数腺体，侧脉密，边脉极接近叶缘；叶柄长1～2mm。聚伞花序顶生，长1～1.5cm，少花；花梗长3～4mm，花白色；萼管长2mm，萼齿极短；花瓣4，分离，近圆形，长约2mm；雄蕊多数；子房下位，花柱与雄蕊同长。果实球形，径4～5mm。花期5～6月。

产宜章。生山坡灌丛、山坡疏林、山谷或溪边。分布于浙江南部、福建、广东、广西、江西、贵州。

Evergreen shrubs. Young twigs slender, 4-angled, after dry dark brown. Leaves leathery, small, usually ternate, narrowly oblong or narrowly lanceolate, 1.5-2cm long, apex obtuse or slightly acute, base cuneate, after dry dark brown adaxially, with numerous glands abaxially, lateral veins dense, intramarginal veins very close to margin; petioles 1-2mm long. Cymes terminal, 1-1.5cm long, few flowered; pedicels 3-4mm long, flowers white; calyx tube 2mm long, calyx teeth short; petals 4, free, suborbicular, ca.2mm long; stamens numerous; ovary inferior, style as long as stamens. Fruit globose, 4-5mm in diam.. Fl. May-Jun.

Distributed in Hunan(Yizhang), South Zhejiang, Fujian, Guangdong, Guangxi, Jiangxi and Guizhou. Grows in thickets and sparse forests on mountain slopes, and in valleys or along ravines.

常绿灌木或小乔木。嫩枝圆柱形，干后黑褐色。叶革质，狭椭圆形至长圆形或倒卵形，长3～7cm，先端钝或略尖，基部宽楔形至楔形，边缘稍背卷，上面有光泽，具多数细小而下陷的腺体，侧脉密，不明显，在距叶缘约0.5mm处连结成边脉；叶柄长3～6mm。圆锥花序腋生和顶生，长1～2cm；花白色，无花梗，常3花簇生花序轴顶端；花芽倒卵形，长2mm，萼管倒圆锥形，长1.5mm，萼齿不明显；花瓣4，分离，圆形，长1mm。果实球形，径5～6mm。花期7～9月。

产通道。生低山林中。分布于福建、广东、广西、海南。树形雅致，枝繁叶茂，终年翠绿，嫩枝嫩叶鲜红色，为优良的庭园绿化、观赏树种。

Evergreen shrubs or small trees. Young twigs terete, after dry black-brown. Leaves leathery, narrowly elliptic to oblong or obovate, 3-7cm long, apex obtuse or slightly acute, base broadly cuneate to cuneate, margin slightly reflexed, glossy adaxially, with many small and impressed glands, lateral veins dense, inconspicuous, intramarginal veins ca.0.5 mm from margin; petioles 3-6mm long. Panicles axillary and terminal, 1-2cm long; flowers white, sessile, usually 3 flowers clustered at the top of inflorescence rachis；flower buds obovate, 2mm long; calyx tube obconic, 1.5mm long, calyx teeth inconspicuous; petals 4, free, rounded, 1mm long. Fruit globose, 5-6mm in diam.. Fl. Jul-Sep.

Distributed in Hunan(Tongdao), Fujian, Guangdong, Guangxi and Hainan. Grows in low mountain forests. It is an excellent garden and ornamental tree for its elegant tree-shape, dense twigs and leaves, year-round green leaves, and young shiny red leaves.

常绿乔木。嫩枝圆柱形，红色，老枝灰褐色。叶革质，对生，椭圆形至狭椭圆形，长4~7cm，先端尾尖，基部阔楔形，两面无毛，上面干后黑褐色，全缘，侧脉细而多，在距叶缘1~1.5 mm处网结成边脉；叶柄长7~9mm。聚伞花序生于枝顶叶腋，长1~2cm，通常有5~6条分枝，每分枝具无梗的花3朵，花萼倒圆锥形，长约3mm，顶部平截，萼齿不明显。果实椭圆状卵形，长1.5~2cm。花期6~8月。

产通道、江永。生低海拔山坡疏林、山谷、沟边、常绿阔叶林中。分布于福建、广东、广西、四川。幼叶鲜红，可引种栽培为观赏树种。

Evergreen trees. Young twigs terete, red, old branches gray brown. Leaves leathery, opposite, elliptic to narrowly elliptic, 4-7cm long, apex caudate, base broadly cuneate, glabrous on both surfaces, after dry dark brown adaxially, margin entire, lateral veins thin and numerous, intramarginal veins 1-1.5 mm from margin; petioles 7-9mm long. Cymes axillary in axils apically on branches, 1-2cm long, usually with 5-6 branches, each branch with sessile flowers 3, calyx obconic, ca.3mm long, apically flatted truncate, calyx teeth inconspicuous. Fruit elliptic-ovate, 1.5-2cm long. Fl. Jun-Aug.

Distributed in Hunan(Tongdao, Jiangyong), Fujian, Guangdong, Sichuan and Guangxi. Grows in sparse forests on mountain slopes, valleys, along ravines, evergreen broad-leaved forests at low altitude. It can be introduced for cultivation ornamental tree for its young red leaves.

线萼金花树（黄金梢） *Blastus apricus* (H.-M.) H. L. Li

常绿灌木。茎圆柱形，幼枝、叶柄、叶下面、花各部均被微柔毛及黄色腺体。叶纸质，披针形至卵状披针形或卵形，长4~14cm，先端渐尖，基部圆形或微心形，全缘或具细波状齿，基出5脉，上面无毛，下面散生黄色腺体；叶柄长3~20mm。圆锥花序顶生，长4~10cm；花萼漏斗形，具四棱，长约5mm，裂片线状三角形，长1~1.5mm；花瓣紫红色，长约2.5mm；雄蕊4，等长。蒴果椭圆形，包于宿存萼中，长约5mm。

花期6~7月，果期10~11月。

产城步、洞口、江华、通道、绥宁、蓝山、汝城、炎陵。生海拔300~800m山谷、山坡林下。分布于福建、广东、广西、江西。

Evergreen shrubs. Stem terete, young twigs, petioles, lower surface of leaves, various parts of flowers puberulous and yellow glands. Leaves papery, ovate or lanceolate to ovate-lanceolate, 4-14cm long, apex acuminate, base rounded or slightly cordate, margin entire or finely undulate-dentate, basal veins 5, glabrous adaxially, scattered yellow glands abaxially; petioles 3-20mm long. Panicles terminal, 4-10cm long; calyx infundibulate, 4-angled, ca.5mm long, lobes linear triangle, 1-1.5mm long; petals purplish-red, ca.2.5mm long; stamens 4, equal in length. Capsule elliptic, wrapped in persistent calyx, ca.5mm long. Fl. Jun-Jul, fr. Oct-Nov. Distributed in Hunan (Chengbu, Dongkou, Jianghua, Tongdao, Suining, Lanshan, Rucheng, Yanling), Fujian, Guangdong, Guangxi and Jiangxi. Grows in valleys and in forests on mountain slopes at alt. 300-800m.

匙萼柏拉木　　*Blastus cavaleriei* H. Lév. et Vaniot

常绿灌木。茎圆柱形，幼枝、花序梗、花梗、花萼均密被锈色微柔毛和稀疏腺体。叶纸质，卵形或披针状卵形，长6.5～14cm，先端渐尖，基部心形至圆形，全缘或具细锯齿，基出5脉，上面近无毛，下面仅脉上被锈色微柔毛及腺体；叶柄长0.7～2.3cm。圆锥花序顶生，长4.5～9cm；花萼漏斗形，具4棱，长4～5mm，裂片匙形，长2～3mm，顶端圆；花瓣紫红色，长圆形，长约5mm；雄蕊4，等长。蒴果椭圆形，包于宿存花萼中，长约4mm，被腺体。花期6～8月，果期8～11月。

产通道、东安、宁远、城步、宜章。生山谷、山坡林下。分布于广东、广西、贵州、云南。叶捣碎外敷，可止血。

Evergreen shrubs. Stem terete, young twigs, peduncle, pedicels, calyx densely rusty puberulous and sparsely glands. Leaves papery, ovate or lanceolate-ovate, 6.5-14cm long, apex acuminate, base cordate to rounded, margin entire or minutely serrate, basal veins 5, subglabrous adaxially, only veins rusty puberulous and glands abaxially; petioles 0.7-2.3cm long. Panicles terminal, 4.5-9cm long; calyx infundibulate, 4-angled, 4-5mm long, lobes spatulate, 2-3mm long, apex rounded; petals purple red, oblong, ca.5mm long; stamens 4, equal in length. Capsule elliptic, wrapped in persistent calyx, ca.4mm long, glands. Fl. Jun-Aug, fr. Aug-Nov.

Distributed in Hunan (Tongdao, Dongan, Ningyuan, Chengbu, Yizhang), Guangdong, Guangxi, Guizhou and Yunnan. Grows in valleys and in forests on mountain slopes. Leaves can stop bleeding by pounding and applying on the outside.

柏拉木

Blastus cochinchinensis Lour.

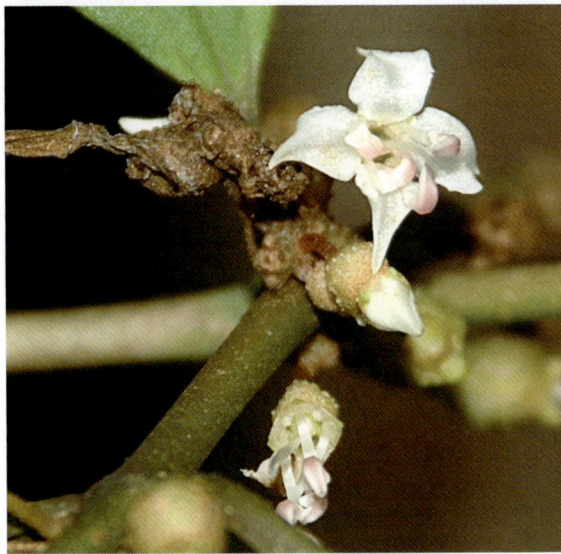

常绿灌木。小枝圆柱形,幼枝被黄褐色腺体。叶对生,披针形至椭圆状披针形,长6～12cm,先端渐尖,基部楔形,全缘或具小波状齿,基出3(5)脉,两面无毛,下面密被腺体;叶柄长1～2cm。聚伞花序腋生,总花梗和花梗均被腺体;花萼钟状,长约4mm,被腺体,顶端4裂;花瓣4,白色,稀粉红色,长约4mm;雄蕊4,等长。蒴果椭圆形,4纵裂,宿萼与果近等长,被腺体。花期6～8月,果期10～12月。

产通道。生海拔200m溪边。分布于福建、广东、广西、海南、台湾、云南。印度及越南亦产。

Evergreen shrubs. Branchlets terete, young twigs yellow-brown glands. Leaves opposite, lanceolate to elliptic-lanceolate, 6-12cm long, apex acuminate, base cuneate, margin entire or finely undulate-dentate, basal veins 3(5), glabrous on both surfaces, densely glands abaxially; petioles 1-2cm long. Cymes axillary, peduncle and pedicels glands; calyx campanulate, ca.4mm long, glands, apex 4-lobed; petals 4, white, rarely pink, ca.4mm long; stamens 4, equal in length. Capsule elliptic, 4-valved, persistent calyx and fruit subequal in length, glandular. Fl. Jun-Aug, fr. Oct-Dec.

Distributed in Hunan(Tongdao), Fujian, Guangdong, Guangxi, Hainan, Taiwan and Yunnan, also in India, Vietnam. Grows in streamsides at alt. 200m.

金花树   *Blastus dunnianus* Levl.

常绿灌木。幼枝、叶柄、花序各部均被锈色微柔毛和黄色腺体。叶对生，卵状披针形、椭圆状披针形至卵形，长8～15cm，先端渐尖，基部浅心形，全缘或具细波状齿，基出5脉，上面无毛，下面沿脉密被黄色腺体；叶柄长8～20mm。圆锥花序顶生；花红色；萼筒长约3～5mm，裂片4，反折，椭圆形，顶端微凹或钝，长约2mm；花瓣4，长2～3mm；雄蕊4，等长。蒴果椭圆形，4纵裂。花期6～7月，果期9～11月。

产新宁、城步、绥宁、通道、道县、宁远、桂东、宜章、资兴。生山坡林下。分布于福建、广东、广西、贵州、江西。全株入药，治风湿及止血。

Evergreen shrubs. Young twigs, petioles, various parts of inflorescences rusty puberulous and yellow glands. Leaves opposite, ovate-lanceolate, elliptic-lanceolate to ovate, 8-15cm long, apex acuminate, base shallowly cordate, margin entire or finely undulate-dentate, basal veins 5, glabrous adaxially, densely yellow glands on veins abaxially; petioles 8-20mm long. Panicles terminal; flowers red; calyx tube ca.3-5mm long, lobes 4, reflexed, elliptic, apex retuse or obtuse, ca.2mm long; petals 4, 2-3mm long; stamens 4, equal in length. Capsule elliptic, 4-valved. Fl. Jun-Jul, fr. Sep-Nov.

Distributed in Hunan(Xinning, Chengbu, Suining, Tongdao, Daoxian, Guidong, Ningyuan, Yizhang, Zixing), Fujian, Guangdong, Guangxi, Guizhou, and Jiangxi. Grows in forests on mountain slopes. Whole plant can be used as medicine, rheumatism, and hemostatic.

常绿灌木。幼枝、叶柄、叶下面、花各部均被微柔毛及黄色腺体。叶对生，卵状披针形至椭圆状披针形，长3～7cm，先端渐尖，基部圆形或浅心形，全缘或具细齿，基出5脉，上面无毛，下面密被黄色腺体；叶柄长0.5～1cm。圆锥花序顶生；花两性，紫红色；萼筒长2～3mm，顶端4裂，裂片小，三角形；花瓣4，长2～2.5mm；雄蕊4，等长。蒴果椭圆形，4纵裂。花期7月，果期10月。

产通道、道县、城步、宜章、新宁、资兴。生低海拔山谷林下、溪边。分布于江西南部、广东。

Evergreen shrubs. Young twigs, petioles, lower surface of leaves, various parts of flowers puberulous and yellow glands. Leaves opposite, ovate lanceolate to elliptic-lanceolate, 3-7cm long, apex acuminate, base rounded or shallowly cordate, margin entire or minutely denticulate, basal veins 5, glabrous adaxially, densely yellow glands abaxially; petioles 0.5-1cm long. Panicles terminal; flowers bisexual, purplish-red; calyx tube 2-3mm long, top 4-lobed, lobes small, triangular; petals 4, 2-2.5mm long; stamens 4, equal in length. Capsule elliptic, 4-valved. Fl. Jul, fr. Oct.

Distributed in Hunan(Tongdao, Daoxian, Chengbu, Yizhang, Xinning, Zixing), South Jiangxi and Guangdong. Grows in valley forests, and streamsides at low altitude.

常绿灌木。小枝略具棱，密被伏贴的鳞片状糙伏毛。叶对生，披针形或卵状披针形，长6～13cm，先端渐尖，基部近圆形或近楔形，基出5脉，两面密被紧贴糙伏毛；叶柄5～10cm，密被糙伏毛。伞房花序具2～7花，生于枝顶；花梗长5～8mm；花序梗、花梗、花萼均被糙伏毛。花粉红色至红色；萼筒长8～10mm，裂片5；花瓣5，倒卵形；雄蕊10，5长5短。果实坛状球形，顶端平截，稍肉质，不开裂，径6～8mm，外面密被鳞片状糙伏毛。花期2～5月，果期8～12月。

产通道南部、江永。生海拔300m以下山谷及林下。分布于四川、云南、广西、广东、福建、台湾。印度、泰国、马来西亚、澳大利亚亦产。果可食。

Evergreen shrubs. Branchlets slightly angled, densely appressed squamose strigose. Leaves opposite, lanceolate or ovate-lanceolate, 6-13cm long, apex acuminate, base subrounded or subcuneate, basal veins 5, densely appressed strigose on both surfaces; petioles 5-10cm long, densely strigose. Corymbs terminal, with 2-7 flowers; pedicels 5-8mm long; peduncle, pedicels, and sepals strigose. Flowers pink to red; calyx tube 8-10mm long, lobes 5; petals 5, obovate; stamens 10, 5 long and 5 short. Fruit urn-shaped globose, apically flatted truncate, slightly fleshy, indehiscent, 6-8mm in diam., outside densely squamose strigose. Fl. Feb-May, fr. Aug-Dec.

Distributed in Hunan (South Tongdao, Jiangyong), Sichuan, Yunnan, Guangxi, Guangdong, Fujian and Taiwan, also in India, Thailand, Malaysia, Australia. Grows in valleys and forests below alt. 300m. Fruit is edible.

常绿灌木。茎钝四棱形或近圆柱形，密被紧贴的鳞片状毛。叶对生，卵形或宽卵形，长4～10cm，先端急尖，基部浅心形或圆形，全缘，基出7脉，两面密被糙伏毛及短柔毛；叶柄长5～15mm，密被鳞片状糙伏毛。伞房花序生枝顶，具1～5花；萼筒长约8～10mm，密生伏贴的鳞片状毛，裂片5；花瓣倒卵形，玫瑰红色；花瓣5；雄蕊10，5长5短。果实坛状球形，稍肉质，不开裂，长约1～1.5cm，外面密被伏贴的鳞片状毛。花期5～7月，果期10～12月。

产通道南部。生海拔200～400m的马尾松林下。湖南南岭地区稀见。分布于广西、广东、福建、台湾。中南半岛亦产。

Evergreen shrubs. Stem obtusely 4-angled or subterete, densely appressed scalelike hairy. Leaves opposite, ovate or broadly ovate, 4-10cm long, apex acute, base shallowly cordate or rounded, margin entire, basal veins 7, densely strigose and pubescent on both surfaces; petioles 5-15mm long, densely squamose strigose. Corymbs terminal, with 1-5 flowers; calyx tube ca.8-10mm long, densely appressed scalelike hairy, lobes 5; petals obovate, rose red; petals 5; stamens 10, 5 long and 5 short. Fruit urn-shaped globose, slightly fleshy, indehiscent, ca.1-1.5cm long, outside densely appressed scalelike hairs. Fl. May-Jul, fr. Oct-Dec.

Distributed in Hunan (South Tongdao), Guangxi, Guangdong, Fujian and Taiwan, also in Indo-China peninsula. Grows in *Pinus massoniana* forests at alt. 200-400m. It is a rare species in Nanling region of Hunan.

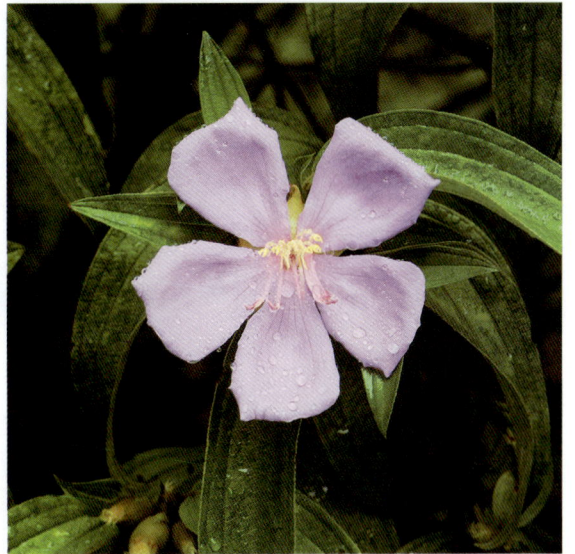

展毛野牡丹　　*Melastoma normale* D. Don

常绿灌木。小枝密被展开的长粗毛及糙伏毛，略具棱。叶对生，椭圆形至狭椭圆状披针形，长4～12cm，先端渐尖，基部楔形或圆形，两面密被糙伏毛，基出5脉；叶柄长5～15mm，密被糙伏毛。伞房花序具3～10花，生于枝顶；花梗2～5mm；花序及花梗密被糙伏毛；萼筒长6～8mm，密生糙伏毛，裂片5；花瓣5，紫红色；雄蕊10，5长5短。蒴果坛状或近球形，稍肉质，不开裂，径约6～10mm，密生伏贴的鳞片状毛。花期春至夏初，果期秋季。

产通道南部。生海拔300m马尾松疏林下。分布于福建、广东、四川、台湾、西藏、云南、广西、海南。印度、泰国、印度尼西亚、菲律宾亦产。果可食。

Evergreen shrubs. Branchlets densely spread long coarse hair and strigose, slightly angled. Leaves opposite, oblong to narrowly elliptic-lanceolate, 4-12cm long, apex acuminate, base cuneate or rounded, densely strigose on both surfaces, basal veins 5; petioles 5-15mm long, densely strigose. Corymbs terminal, with 3-10 flowers; pedicels 2-5mm long; inflorescences and pedicels densely strigose; calyx tube 6-8mm long, densely strigose, lobes 5; petals 5, red-purple; stamens 10, 5 long and 5 short. Capsule urn-shaped or subglobose, slightly fleshy, indehiscent, ca.6-10mm in diam., densely appressed scalelike hairs. Fl. Spring to early Summer, fr. Autumn.

Distributed in Hunan (South Tongdao), Fujian, Guangdong, Sichuan, Taiwan, Tibet, Yunnan, Guangxi and Hainan, also in India, Thailand, Indonesia, Philippines. Grows in *Pinus massoniana* sparse forests at alt. 300m. Fruit is edible.

　　小灌木。茎四棱形或稀六棱形，被平贴糙伏毛或上升糙伏毛。叶坚纸质，对生或稀3枚轮生，卵形至卵状披针形，长5.5～11.5cm，先端渐尖，基部钝或圆形，基出5脉，全缘，具缘毛，两面密被糙伏毛；叶柄长0.5～1cm，密被平贴糙伏毛。圆锥花序顶生；花萼裂片4，卵状三角形，外被刺形星状毛和微柔毛；花瓣4，深红色至紫色，卵形，长约2cm；雄蕊8，等大，偏于一侧。蒴果长卵圆形，宿存萼长坛状，近中部缢缩成颈，外被刺形星状毛。花果期7～9月。

　　产城步、通道、江华、江永。生低海拔草坡、路边、疏林下。分布于浙江南部、福建、台湾、江西、广东、广西、海南、贵州、四川南部。越南至泰国亦产。

Small shrubs. Stem 4-angled or rarely 6-angled, densely appressed strigose or ascending strigose. Leaves hard papery, opposite or rarely verticillate, ovate to ovate-lanceolate, 5.5-11.5cm long, apex acuminate, base obtuse or rounded, basal veins 5, margin entire, ciliate, densely strigose on both surfaces; petioles 0.5-1cm long, densely appressed strigose. Panicles terminal; calyx lobes 4, ovate-triangular, outside spinous stellate-pubescent and puberulous; petals 4, dark red to purple, ovate, ca.2cm long; stamens 8, equal in length, partial to one side. Capsule long ovoid, persistent calyx long urn-shaped, nearly middle constricted into the neck, outside densely spinous stellate-pubescent. Fl. and fr. Jul-Sep.

Distributed in Hunan (Chengbu, Tongdao, Jianghua, Jiangyong), South Zhejiang, Fujian, Taiwan, Jiangxi, Guangdong, Guangxi, Hainan, Guizhou and South Sichuan, also in Vietnam to Thailand. Grows in low grassy slopes, roadsides, sparse forests at low altitude.

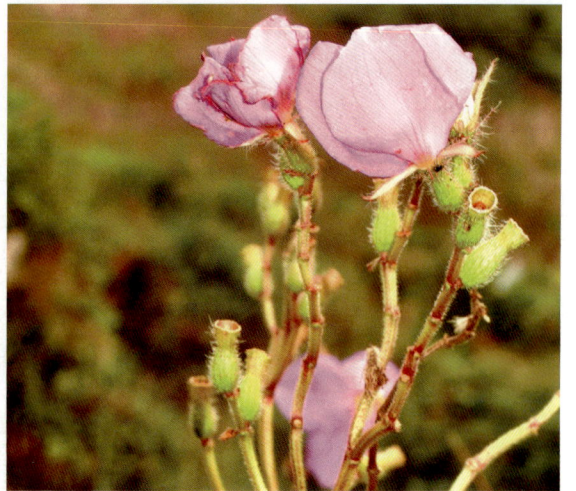

假朝天罐 *Osbeckia crinita* Benth. ex C. B. Clark

灌木。枝四棱形,被弯曲或伸展的粗毛。叶坚纸质,对生,长圆状披针形,长5~9cm,基部钝或近心形,基出5脉,全缘,具缘毛,上面被糙伏毛,下面仅脉上被糙伏毛;叶柄长3~10mm,密被糙伏毛。圆锥花序顶生;苞片2,卵形;花梗短或几无,花萼裂片4,线状披针形或钻形;花瓣4,紫红色,倒卵形,长约1.5cm,具缘毛;雄蕊8,等大,偏于一侧。蒴果卵圆形,宿存萼长坛状,近中部缢缩成颈,外密被刺形星状毛。花期8~11月,果期10~12月。

产黔阳、新宁、城步、宜章、江华、江永、道县、通道、洞口、武冈、保靖、会同、桑植、石门。生低海拔草坡、路边、疏林下。分布于湖北、福建、广东、广西、贵州、四川、云南、西藏。

Shrubs. Branchlets 4-angled, crooked or outstretched coarse hair. Leaves hard papery, opposite, oblong-lanceolate, 5-9cm long, base obtuse or subcordate, basal veins 5, margin entire, ciliate, strigose adaxially, only veins strigose abaxially; petioles 3-10mm long, dense strigose. Panicles terminal; bracts 2, ovate, ciliate; pedicels short or absent, calyx lobes 4, linear-lanceolate or subulate; petals 4, red-purple, obovate, ca.1.5cm long, ciliate; stamens 8, equal in length, partial to one side. Capsule ovate-orbicular, persistent calyx long urn-shaped, nearly middle constricted into the neck, outside densely spinous stellate-pubescent. Fl. Aug-Nov, fr. Oct-Dec.

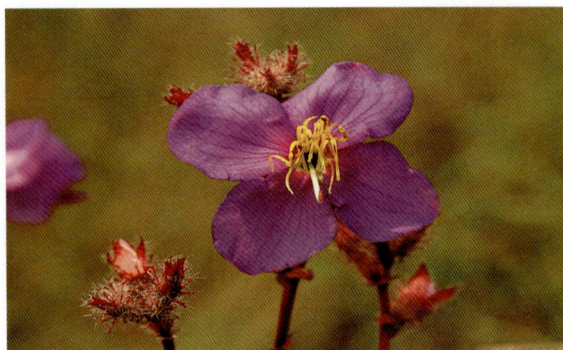

Distributed in Hunan(Qianyang, Xinning, Chengbu, Yizhang, Jianghua, Jiangyong, Daoxian, Tongdao, Dongkou, Wugang, Baojing, Huitong, Sangzhi, Shimen), Hubei, Fujian, Guangxi, Guizhou, Guangdong, Sichuan, Yunnan and Tibet. Grows in grassy slopes, roadsides, sparse forests at low altitude.

锦香草（熊巴掌） *Phyllagathis cavaleriei* (Levl. et Vant.) Guillanum

多年生草本。茎直立或匍匐，密被长粗毛，四棱形，常无分枝。叶对生，整株仅2～3对，宽卵形或圆形，长4.5～9cm，先端急尖或近圆，基部心形，边缘不明显细波状，具缘毛，基出7脉，上面疏被伏贴长粗毛，下面仅脉上被平展长粗毛；叶柄长1.5～9cm，密被长粗毛。伞形花序顶生，花序梗近无毛；花粉红色至紫色；萼筒4棱，近无毛，裂片4，宽卵形；花瓣4，宽倒卵形；雄蕊8，等大；蒴果杯形，径约6mm。

产城步、通道、东安、溆浦、黔阳、新宁、江华、桂东、宜章。生山地林下、水湿地。分布于广西、广东、福建、江西、贵州。叶形美观，花红色，可作盆栽观赏。

Perennial herbs. Stem erect or prostrate, densely long thick trichomes, 4-angled, usually unbranched. Leaves opposite, only 2-3 pairs, broadly ovate or rounded, 4.5-9cm long, apex acute or nearly rounded, base cordate, margin inconspicuous minutely repand and ciliate, basal veins 7, sparsely appressed long thick trichomes adaxially, only veins spreading thick trichomes abaxially; petioles 1.5-9cm long, densely long thick trichomes. Umbels terminal, peduncle subglabrous; flowers pink to purple; calyx tube 4-angled, subglabrous, lobes 4, broadly ovate; petals 4, broadly obovate; stamens 8, equal in length. Capsule cupular, ca.6mm long.

Distributed in Hunan (Chengbu, Tongdao, Dongan, Xupu, Qianyang, Xinning, Jianghua, Guidong, Yizhang), Guangxi, Guangdong, Fujian, Jiangxi and Guizhou. Grows in forests, wetland in mountain land. It can be used as ornamental plants for its beautiful and red flowers.

使君子　　*Quisqualis indica* L.

　　落叶攀援藤本。嫩枝、幼叶被黄褐色短柔毛。叶薄纸质，对生，椭圆形至卵形，长5～13cm，先端渐尖，基部钝圆，全缘至波状，侧脉7～8对；叶柄长5～10mm，幼时密被柔毛，叶柄下部宿存成硬刺状，穗状花序顶生，下垂；萼筒纤细，绿色，长达9cm，顶端5裂，被柔毛；花瓣5，长圆形，长1.5～2cm，初为白色，后变为淡红色；雄蕊10，2轮排列。果卵状纺锤形，长2.5～4cm，具5锐棱，熟时黑色，顶部开裂3～5瓣。

　　产江华、江永、宜章。生海拔200～500m低山荒地、路边。分布于福建、广东、广西、贵州、江西、四川、台湾、云南。印度、缅甸、菲律宾亦产。种子入药可以毒杀肠道寄生虫。花美丽，常栽观赏。

　　Deciduous climbing vines. Young twigs, young leaves yellow-brown pubescent. Leaves thinly papery, opposite, oblong to ovate, 5-13cm long, apex acuminate, base obtuse, margin entire or undulate, lateral veins 7-8 pairs; petioles 5-10mm long, densely pilose when young, its lower part hard spiny. Spikes terminal, pendulous; calyx tube slender, up to 9cm long, apex 5-lobed, pilose; petals 5, oblong, 1.5-2cm long, white at first, later becoming pale red; stamens 10, in 2 series. Fruit ovate-fusiform, 2.5-4cm long, sharply 5-ridged, black when mature, apex 3-5-valved.

　　Distributed in Hunan (Jianghua, Jiangyong, Yizhang), Fujian, Guangdong, Guangxi, Guizhou, Jiangxi, Taiwan, Yunnan and Sichuan, also in India, Burma, and Philippines. Grows in low mountain and moorland, roadsides at alt. 200-500m. The seeds are used medicinally to kill intestinal parasites. This species is usually cultivated as an ornamental for its beautiful flowers.

多花山竹子 *Garcinia multiflora* Champ. ex Benth.

常绿乔木。小枝光绿色。叶革质，对生，倒卵状矩圆形，长7～15cm，先端钝或突尖，基部楔形，全缘，两面无毛，中脉在上面微凸起，侧脉纤细，多而密；叶柄长1～2cm。花杂性同株，雄花排成圆锥花序；总花梗和花梗具关节；花橙黄色，花丝合生4束；雌花序1～5花；柱头无柄，盾形，大而厚。浆果近球形，长4～5cm，黄绿色，光滑，顶端具宿存的柱头。花期7～8月，果期11～12月。

产通道、城步、绥宁、江华、江永、洞口、汝城、炎陵、资兴。生海拔700m以下溪边、沟谷、阔叶林中。分布于江西、台湾、福建、广东、广西、贵州、云南、海南。越南亦产。果酸甜可食，但不宜多食；树冠浓密，四季常青，可引种为园林绿化树。

Evergreen trees. Branchlets light green. Leaves leathery, opposite, obovate-oblong, 7-15cm long, apex obtuse or abruptly acuminate, base cuneate, margin entire, glabrous on both surfaces, midvein slightly raised adaxially, lateral veins slender, numerous and dense; petioles 1-2cm long. Flower polygamo-monoecious, male flowers arranged in panicles; peduncle and pedicels with joint; flowers yellow orange, filaments connate into 4 bundles; female inflorescences with 1-5 flowers; stigma sessile, peltate, large and thick. Berry subglobose, 4-5cm long, yellow green, smooth, apex with persistent stigma. Fl. Jun-Aug, fr. Nov-Dec.

Distributed in Hunan(Tongdao, Chengbu, Suining, Jianghua, Jiangyong, Dongkou, Rucheng, Yanling, Zixing), Jiangxi, Taiwan, Fujian, Guangdong, Guangxi, Guizhou, Yunnan and Hainan, also in Vietnam. Grows in streamsides, ravines, broad-leaved forests below alt. 700m. Fruit is sweet and edible, but not proper for eating too much. It can be introduced as a landscape tree for its bushy and year-round evergreen canopy.

山芝麻　*Helicteres angustifolia* L.

落叶小灌木。小枝被灰黄色短柔毛。叶条状披针形，长3～8cm，先端钝或短尖，基部圆形，全缘，下面被灰白色或淡黄色星状短柔毛，基出3脉；叶柄长3～8mm，密被柔毛及星状毛。聚伞花序顶生或腋生；花萼管状，5浅裂，裂片三角形；花瓣5，不等长，红色或紫红色。蒴果卵状长圆形，长约1.5cm，先端短尖，密被星状毛。花期4～10月。

产江华、江永、宜章。生海拔300m以下荒地草坡。分布于福建、广东、广西、江西、台湾、云南、海南。东南亚各国亦产。茎皮纤维为优质纤维原料。

Deciduous small shrubs. Branchlets grayish-yellow pubescent. Leaf strip-lanceolate, 3-8cm long, apex obtuse or mucronate, base rounded, margin entire, gray white or pale yellow stellate-pubescent abaxially, basal veins 3; petioles 3-8mm long, densely pilose and stellate-pubescent. Cymes terminal or axillary; calyx tube-shaped, 5-lobed, lobes triangular; petals 5, unequal in length, red or purple-red. Capsule ovate-oblong, ca.1.5cm long, apex mucronate, densely stellate-pubescent. Fl. Apr-Oct.

Distributed in Hunan (Jianghua, Jiangyong, Yizhang), Fujian, Guangdong, Guangxi, Jiangxi, Taiwan, Yunnan, Hainan, also in Southeast Asian countries. Grows in wasteland and grassy slopes below alt. 300m. The bark fiber is a high-quality fiber material.

常绿乔木。小枝密被黄褐色柔毛。叶2形，生于幼树或萌枝上的叶盾形，掌状3～5裂，下面密被褐色柔毛；正常枝条及花果枝上的叶长圆状卵形或矩圆形，长7～15cm，先端渐尖或钝，基部斜圆形或斜心形，下面密被黄褐色茸毛，基出3～5脉，侧脉7～8对。花序腋生，具1～4朵花，萼片5，线形，两面密被毛；花瓣5，倒披针形，白色。蒴果木质，狭卵形，长约6cm，密被星状柔毛；种子具膜质翅。

产通道、新宁、江华、江永。生海拔500m以下沟谷林中或石灰岩山地。分布于福建、广东、广西、海南。喜光，速生，干直，为优良用材及纤维树种。

Evergreen trees. Branchlets densely yellow-brown villous. Leaves dimorphic, peltate at young tree and germination branches, palmately 3-5-cleft, densely brown villous; oblong-ovate or oblong at normal branches, floral branches and fruit branches, 7-15cm long, apex acuminate or obtuse, base obliquely rounded or obliquely cordate, densely yellow-brown tomentose abaxially, basal veins 3-5, lateral veins 7-8 pairs. Inflorescences axillary, with 1-4 flowers; sepals 5, linear, densely hairy on both surfaces; petals 5, oblanceolate, white. Capsule woody, narrowly ovate, ca.6cm long, densely stellate-pilose; seeds with membranous wings.

Distributed in Hunan (Tongdao, Xinning, Jianghua, Jiangyong), Fujian, Guangdong, Guangxi and Hainan. Grows in ravine forests or in limestone mountain land below alt. 500m. It is a fine material and fiber tree for liking sun, fast-growing and straight trunk.

白背黄花稔 *Sida rhombifolia* L.

　　直立亚灌木。小枝常呈红色，嫩叶、叶柄及花梗均密被灰色短茸毛。叶菱形或长圆状披针形，长2.5～4.5cm，先端钝或急尖，基部宽楔形至圆钝，边缘具锯齿，上面近无毛，下面密被灰色星状茸毛；叶柄长3～5mm；托叶线形，与叶柄近等长。花单生叶腋，花梗长1～2cm，中部以上具关节；花萼钟状，外被短柔毛，裂片5，三角形；花冠黄色，花瓣倒卵形，先端圆，基部狭。果半球形，分果爿8～10，疏被星状柔毛，顶端具2短芒。花期5～12月。

　　产江华、江永、通道。生低海拔山坡灌丛、旷野。分布于台湾、福建、广东、广西、贵州、云南、四川、湖北。中南半岛、印度亦产。

Erect subshrubs. Branchlets usually red, young leaves, petioles and pedicels densely gray tomentellate. Leaves rhombic or oblong-lanceolate, 2.5-4.5cm long, apex obtuse or acute, base broadly cuneate to rounded, margin serrate, subglabrous adaxially, densely gray stellate tomentose abaxially; petioles 3-5mm long; stipules linear, and with petioles subequal. Flowers solitary in leaf axils, pedicels 1-2cm long, with joint above middle; calyx campanulate, outside pubescent, lobes 5, triangular; corolla yellow, petals obovate, apex rounded, base narrowly. Fruit semi-globose, mericarps 8-10, sparsely stellate pilose, apex with 2 short awns. Fl. May-Dec.

　　Distributed in Hunan (Jianghua, Jiangyong, Tongdao), Taiwan, Fujian, Guangdong, Guangxi, Guizhou, Yunnan, Sichuan and Hubei, also in South Peninsula, India. Grows in thickets on mountain slopes and wilderness at low altitude.

梵天花 *Urena procumbens* L.

落叶小灌木。小枝密被星状柔毛。叶互生，形状多变，下部的圆卵形，上部的菱状卵形或卵形，长2～7cm，常3～5浅裂至中部，有时不分裂，基出3～5脉，边缘具钝锯齿，两面密生星状柔毛，下面淡灰绿色；叶柄长4～18mm。花1～2朵生叶腋，淡红色，径约1.5cm；小苞片5，基部合生；花萼钟状，5裂；花瓣5，倒卵形。果扁球形，径约1cm，分果爿5，被钩状刺毛。花期6～9月。

产新宁、城步、江华、江永、宜章、资兴。生海拔500m以下山坡灌丛、路边。分布于广西、广东、江西、福建、台湾、浙江南部。茎皮纤维可编绳。

Deciduous small shrubs. Branchlets densely stellate pilose. Leaves alternate, shape changeable, leaf blades on lower part of stem round-elliptic, leaf blades upper part of stem rhombic-ovate or ovate, 2-7cm long, usually 3-5-lobed to middle, sometimes undivided, basal veins 3-5, margin obtusely serrate, densely stellate pilose on both surfaces, glaucous adaxially; petioles 4-18mm long. Flowers axillary, with 1-2 flowers, light red, ca.1.5cm in diam.; bracteoles 5, base connate; calyx campanulate, 5-cleft; petals 5, obovate. Fruit oblate, ca.1cm in diam., mericarps 5, hooked bristles. Fl. Jun-Sep.

Distributed in Hunan (Xinning, Chengbu, Jianghua, Jiangyong, Yizhang, Zixing), Guangxi, Guangdong, Jiangxi, Fujian, Taiwan and South Zhejiang. Grows in thickets on mountain slopes, roadsides below alt. 500m. Bark fiber can be used to compile the rope.

东方古柯 *Erythroxylum sinensis* C. Y. Wu

落叶灌木。全株无毛。小枝红棕色，具明显皮孔。叶纸质，长椭圆形或倒卵形，长2～14cm，先端短渐尖或钝，基部狭楔形，侧脉纤细，不明显；叶柄长2～8mm。花腋生，2～7花簇生叶腋，或单生叶腋，花梗长6～9mm；花萼钟状，5深裂，裂片阔卵形，宿存；花瓣淡红色，卵状长圆形，长3～6mm；雄蕊10，基部合生成浅杯状；子房长圆形，花柱3，分离。核果长圆形，具3棱，红色。花期4～5月，果期7～8月。

产新宁、城步、绥宁、通道、双牌、道县、桂东、宜章。生山地林缘、灌丛。分布于浙江南部、福建、江西、海南、广东、云南、贵州、广西。印度、缅甸亦产。

Deciduous shrubs. Glabrous throughout. Branchlets reddish-brown, conspicuous lenticels. Leaves papery, oblong or obovate, 2-14cm long, apex shortly acuminate or obtuse, base narrowly cuneate, lateral veins slender, inconspicuous; petioles 2-8mm long. Flowers axillary, 2-7 flowers clustered in leaf axils, or solitary in leaf axils, pedicels 6-9mm long; calyx campanulate, deeply 5-lobed, lobes broadly ovate, persistent; petals pale red, obovate-oblong, 3-6mm long; stamens 10, base connate into shallowly cupular; ovary oblong, styles 3, free. Drupe oblong, 3-angled, red. Fl. Apr-May, fr. Jul-Aug.

Distributed in Hunan (Xinning, Chengbu, Suining, Tongdao, Shuangpai, Daoxian, Guidong, Yizhang), South Zhejiang, Jiangxi, Fujian, Yunnan, Hainan, Guizhou, Guangxi and Guangdong, also in India, Burma. Grows in forest margins, thickets in mountain.

五月茶 *Antidesma bunius* (L.) Spreng.

落叶灌木。小枝无毛，被明显皮孔。叶纸质，长椭圆形、倒卵形或长倒卵形，长10～22cm，先端急尖至圆，基部宽楔形或楔形，两面无毛，上面深绿色，常有光泽，侧脉7～11对；叶柄长3～10mm。雄花序穗状，顶生，长6～15cm；雄花无柄，花萼杯状，3～4裂，浅绿色，雄蕊3～4；雌花序总状，顶生，长8～18cm；花盘杯状，肥厚，花萼3～4裂。核果近球形，径8～10mm，肉质，熟时红色。花期5～6月，果期6～11月。

产江华、宜章。生低海拔灌丛中。分布于江西、广东、广西、海南、贵州、云南、西藏。广布亚洲热带至澳大利亚。果实微酸，可食。

Deciduous shrubs. Branchlets glabrous, conspicuous lenticels. Leaves papery, long elliptic, obovate or long obovate, 10-22cm long, apex acute to rounded, base broadly cuneate or cuneate, glabrous on both surfaces, dark green adaxially, usually glossy, lateral veins 7-11 pairs; petioles 3-10mm long. Male inflorescences spicate, terminal, 6-15cm long; male flowers: sessile, calyx cupular, 3-4-cleft, light green, stamens 3-4; female inflorescence racemose, terminal, 8-18cm long; disk cupular, hypertrophy, calyx 3-4-cleft. Drupe subglobose, 8-10mm in diam., fleshy, red when mature. Fl. May-Jun, fr. Jun-Nov.

Distributed in Hunan (Jianghua, Yizhang), Jiangxi, Guangdong, Guangxi, Hainan, Guizhou, Yunnan and Tibet. Widely distributed in tropical form Asia to Australia. Grows in shrubs at low altitude. Fruit is acid and edible.

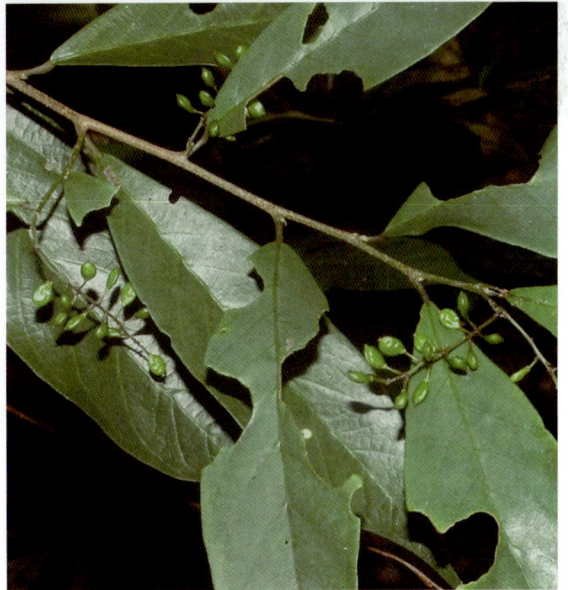

落叶灌木。嫩枝被疏黄色短柔毛。叶纸质，线形或线状披针形，长6～12cm，先端长尾尖，基部楔形，两面光滑无毛，侧脉6～12对，纤细，近平行，在近叶缘处连结；叶柄长2～4mm，被短柔毛。总状花序，有时分枝，顶生或腋生；花序纤细，长1～3cm，被微毛。果序长3～6cm；核果椭圆形，长5～7mm，肉质，熟时红色。花期5～6月，果期6～10月。

产通道、桂阳。生海拔500m以下山地林下。分布于广东、广西、海南、贵州。

Deciduous shrubs. Branchlets sparsely yellow pubescent. Leaves papery, linear or linear-lanceolate, long 6-12cm, apex long acuminate, base cuneate, smooth and glabrous on both surfaces, lateral veins 6-12 pairs, slender, nearly parallel, anastomosing near margin; petioles 2-4mm long, pubescent. Racemes, sometimes branched, terminal or axillary; inflorescences slender, 1-3cm long, puberulous. Infructescence 3-6cm long; drupe elliptic, 5-7mm long, fleshy, red when mature. Fl. May-Jun, fr. Jun-Oct.

Distributed in Hunan (Tongdao, Guiyang), Guangdong, Guangxi, Hainan and Guizhou. Grows in forests in mountain land below alt. 500m.

大叶土蜜树（虾公木） *Bridelia retusa* (L.) A. Jussieu

落叶小乔木。小枝圆柱形，无毛。叶纸质，倒卵形或椭圆形，长8～20cm，先端短急尖或钝，基部圆形，幼嫩时两面被短柔毛，后脱落至无毛，侧脉12～19对，近平行，直达叶缘而网结，网脉明显；叶柄长约1.2cm，被黄褐色柔毛；穗状花序腋生或小枝顶端由3～9个穗状花序再组成圆锥花序状，长10～20cm，被柔毛；花梗长约1mm；萼片被柔毛。核果卵形，长7～8mm，熟时黑色。花期6～7月，果期10～11月。

产江华、江永。生海拔500m以下石灰岩山地。分布于广东、广西、海南、贵州、云南。

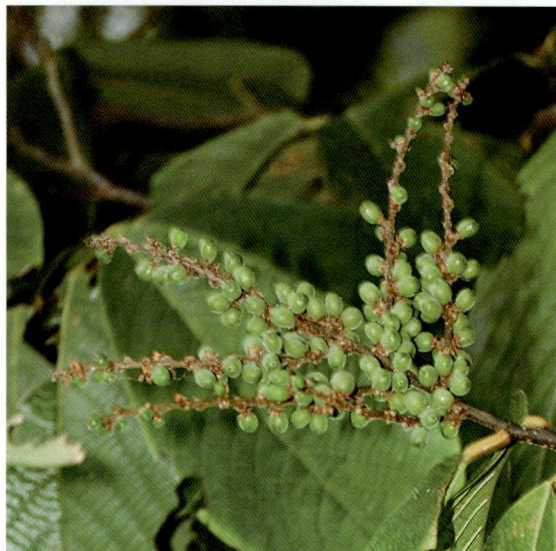

Deciduous small trees. Branchlets terete, glabrous. Leaves papery, obovate or elliptic, 8-20cm long, apex shortly acute or obtused, base rounded, pubescent on both surfaces when young, later gradual glabrous, lateral veins 12-19 pairs, nearly parallel, straight to leaf margin and anastomose, reticulate veins conspicuous; petioles ca.1.2cm long, yellow-brown pilose; inflorescence axillary or 3-9 spikes arranged in paniclelike at apices of branchlets, 10-20cm long, pilose; pedicels ca.1mm long; sepals pubescent. Drupe ovoid, 7-8mm long, black when mature. Fl. Jun-Jul, fr. Oct-Nov.

Distributed in Hunan (Jianghua, Jiangyong), Guangdong, Guangxi, Hainan, Guizhou and Yunnan. Grows in limestone mountain land below alt. 500m.

毛果巴豆 *Croton lachnocarpus* Benth.

落叶灌木。幼枝、幼叶、花序和果均密被星状短柔毛。叶纸质，矩圆形或卵状矩圆形，长4~10cm，先端渐尖或短尖，基部圆形或微心形，边缘具不明显细锯齿，基出3脉，侧脉4~6对，叶基部或顶端有2枚杯状腺体，老叶上面近无毛或仅脉上被星状毛，下面密被星状毛。总状花序顶生，长8~20cm；雄花多数，生花序上部；雌花少数，生花序下部。蒴果扁球形，径6~10mm，被星状毛和长柔毛。花期4~5月。

产新宁、通道、道县、江华、江永、宜章。生海拔600m以下低山林中及石灰岩山地。分布于广东、广西、江西、贵州南部。

Deciduous shrubs. Young twigs, young leaves, inflorescences and fruits densely stellate-pubescent. Leaves papery, oblong or ovate-oblong, 4-10cm long, apex acuminate or acute, base rounded or slightly cordate, margin obscurely serrulate, basal veins 3, lateral veins 4-6 pairs, with 2 cupular glands at leaf base or top, old leaves subglabrous or only veins stellate-pubescent adaxially, densely stellate-pubescent abaxially. Racemes terminal, 8-20cm long; male flowers numerous, in upper part of inflorescence, female flowers few, in lower part of inflorescence. Capsule oblate, 6-10mm in diam., stellate-pubescent and villous. Fl. Apr-May.

Distributed in Hunan(Xinning, Tongdao, Daoxian, Jianghua, Jiangyong, Yizhang), Guangdong, Guangxi, Jiangxi and South Guizhou. Grows in low mountain forests and in limestone mountain below alt. 600m.

巴　豆　　　*Croton tiglium* L.

落叶灌木。幼枝被稀疏星状柔毛，老枝近无毛。叶纸质，卵形或椭圆形，长5～10cm，先端短尖或渐尖，基部宽心形至近圆形，边缘具疏齿，老叶上面近无毛，下面被星状柔毛，基出3(5)脉，侧脉2～3对，基部两侧近叶柄处具2枚盘状腺体；叶柄长2～5cm，近无毛。雌雄同株，总状花序顶生，长8～20cm，雄花生花序轴上部，雌花生下部。蒴果长圆形，长约2cm，近无毛或被短星状毛。花期4～5月，果期7～9月。

产江华、江永、宜章。生海拔200～500m石灰岩山丘。分布于浙江南部、福建、江西、广东、广西、海南、贵州、四川、云南。亚洲南部和东南部各国、菲律宾、日本南部亦产。种子油有剧毒。

Deciduous shrubs. Young twigs sparsely stellate pilose, old branches subglabrous. Leaves papery, ovate or elliptic, 5-10cm long, apex acute or acuminate, base broadly cordate or suborbicular, margin sparsely crenate, old leaves subglabrous adaxially, stellate pilose abaxially, basal veins 3(5), lateral veins 2-3 pairs, with 2 disciform glands on both sides of leaf base; petioles 2-5cm long, subglabrous. Monoecious, racemes terminal, 8-20cm long, female flowers in upper part of inflorescence, female flowers in lower part. Capsule oblong, ca.2cm long, subglabrous or shortly stellate-pubescent. Fl. Apr-May, fr. Jul-Sep.

Distributed in Hunan (Jianghua, Jiangyong, Yizhang), South Zhejiang, Fujian, Jiangxi, Guangdong, Guangxi, Hainan, Guizhou, Sichuan and Yunnan, also in South and Southeast Asia each country, Philippines, South Japan. Grows in limestone hills at alt. 200-500m. The seed oil is very poisonous.

白饭树　*Flueggea virosa* (Willd.) Bailey

常绿灌木。小枝具纵棱槽，全株无毛。叶纸质，椭圆形或长圆形，长2～5cm，先端圆至急尖，基部钝至楔形，全缘或具不整齐波状齿，下面白绿色，侧脉5～8对；叶柄长3～6mm；花小，淡黄色，无花瓣，雌雄异株。雄花簇生叶腋，雄蕊5，退化雌蕊常3深裂；雌花3～10簇生，有时单生，花梗长1.5～12mm，子房卵圆形，3室。蒴果近球形，径3～5mm，熟时淡白色，不开裂。花期3～8月，果期7～12月。

产江永、宜章。生低海拔空旷地及林缘。分布于江西、福建、广东、广西、海南、台湾、贵州、云南。非洲、大洋洲亦产。

Evergreen shrubs. Branchlets with longitudinal furrowed, glabrous throughout. Leaves papery, elliptic or oblong, 2-5cm long, apex rounded or acute, base obtuse to cuneate, margin entire or with irregular undulate-dentate, white-green abaxially, lateral veins 5-8 pairs; petioles 3-6mm long; flowers small, pale yellow, apetalous, dioecious, Male flowers clustered in leaf axils, stamens 5, pistillode usually deeply 3-lobed; female flowers 3-10-fascicled, sometimes solitary, pedicels 1.5-12mm long, ovary ovate-orbicular, 3-loculed. Capsule subglobose, 3-5mm in diam., light white when mature, indehiscent. Fl. Mar-Aug, fr. Jul-Dec.

Distributed in Hunan (Jiangyong, Yizhang), Jiangxi, Fujian, Guangdong, Guangxi, Hainan, Taiwan, Guizhou and Yunnan, also in Africa, Oceania. Grows in empty land and forest margins at low altitude.

东南野桐 *Mallotus lianus* Croiz

落叶小乔木或灌木。树皮红褐色。嫩枝、幼叶、老叶下面、叶柄、花序、花均密被红棕色星状短茸毛。叶纸质，卵圆形或心形，长10～18cm，先端短渐尖，基部钝或近截形，有时稍心形或狭盾形，全缘，下面密生紫红色颗粒状腺体，基出5脉，侧脉5～6对；叶柄长7～13cm。总状花序或圆锥花序；雄花序长10～18cm，每苞腋有3～8朵雄花；雌花序长10～25cm，每苞腋具单花。蒴果扁球形，径8～10mm，密被黄色星状毛和线形软刺。花期8～9月，果期11～12月。

产绥宁、城步、江华、江永、资兴、宜章。生山地灌丛或疏林中。分布于云南、广西、贵州、四川、广东、江西、福建、浙江。

Deciduous small trees or shrubs. Bark reddish-brown. Young twigs, young leaves, lower surfaces of old leaves, petioles, inflorescences, flowers densely red-brown stellate velutinous. Leaves papery, ovate-orbicular or cordate, 10-18cm long, apex shortly acuminate, base obtuse or subtruncate, sometimes slightly cordate or narrowly peltate, margin entire, densely purple granular glands, basal veins 5, lateral veins 5-6 pairs; petioles 7-13cm long. Racemes or panicles; male inflorescences 10-18cm long, each bract axil with 3-8 male flowers; female inflorescences 10-25cm long, each bract axil with 1 flower. Capsule oblate, 8-10mm in diam., densely yellow stellate hairs and linear soft spines. Fl. Aug-Sep, fr. Nov-Dec.

Distributed in Hunan (Suining, Chengbu, Jianghua, Jiangyong, Zixing, Yizhang), Yunnan, Guangxi, Guizhou, Sichuan, Guangdong, Jiangxi, Fujian and Zhejiang. Grows in shrubs or sparse forests in mountain land.

落叶蔓生灌木。幼枝、叶柄、花梗均被微毛。叶纸质，椭圆形或卵形，长1～4cm，先端急尖、钝至圆，基部钝至圆，侧脉5～7对，稍明显；叶柄长2～5cm。雄花2～10朵和雌花1朵簇生叶腋；雄花花梗长5～10mm；萼片5～6；雄蕊5，其中3枚较长，花丝合生，2枚分离；花盘腺体5；雌花花梗长4～8mm；子房4～12室。蒴果浆果状，近球形，径约6mm，红色，干后灰黑色，不开裂。花期3～6月，果期6～10月。

产道县、江永、宜章。生低海拔山地灌丛、林下。分布于江西、福建、台湾、广东、广西、四川、贵州、云南。热带亚洲、大洋洲、非洲亦产。

Deciduous scrambling shrubs. Young branchlets, petioles, and pedicels puberulous. Leaves papery, elliptic or ovate, 1-4cm long, apex acute, obtuse to rounded, base obtuse to rounded, lateral veins 5-7 pairs, slightly conspicuous; petioles 2-5cm long. 2-10 male flowers and 1 female flower fascicled in leaf axils; male flowers: pedicels 5-10mm long; sepals 5-6; stamens 5, 3 longer, filaments connate, 2 free; disk glands 5; female flowers: pedicels 4-8mm long; ovary 4-12-loculed. Capsule berry-like, subglobose, ca.6mm in diam., red, after dry black, indehiscent. Fl. Mar-Jun, fr. Jun-Oct.

Distributed in Hunan(Daoxian, Jiangyong, Yizhang), Jiangxi, Fujian, Taiwan, Guangxi, Guangdong, Sichuan, Guizhou and Yunnan, also in Tropical Asia, Oceania, and Africa. Grows in shrubs and forests at low altitude mountain.

圆叶乌桕　*Sapium rotundifolium* Hemsl.

　　落叶灌木。全株无毛。小枝灰白色。叶近圆形，革质，长5.5～11cm，宽6～12cm，先端圆而具一小凸尖，基部近圆形，上面绿色，下面粉白色，侧脉8～15对；叶柄长3～7cm，顶端有2腺体。总状花序顶生，雌雄同序，无花瓣；雄花5，簇生苞腋内，生上部；雌花少数，生下部。蒴果近球形，径约1.5cm，分果爿木质，脱落；种子近三角形，外被蜡层。花期5～6月，果期9～10月。

　　产道县、江华、江永、蓝山、宁远、宜章、临武。生海拔500m以下石灰岩山丘。分布广东、广西、贵州、云南。树姿优美，秋叶红艳，性喜阳光，为石灰岩山地造林绿化的优良树种。

　　Deciduous shrubs. Glabrous throughout. Branchlets gray white. Leaves suborbicular, leathery, 5.5-11mm × 6-12cm, apex rounded and with a small abruptly mucronate, base subrounded, green adaxially, pinkish-white abaxially, lateral veins 8-15 pairs; petioles 3-7cm long, apex with 2 glands. Racemes terminal, androgynous, apetalous; 5 male flowers clustered in bract axils, in upper part of inflorescence; female flowers few, in lower part of inflorescence. Capsule subglobose, ca.1.5cm in diam., mericarps woody, deciduous; seeds subtriangular, outside with wax layer. Fl. May- Jun, fr. Sep-Oct.

　　Distributed in Hunan (Daoxian, Jianghua, Jiangyong, Lanshan, Ningyuan, Yizhang, Linwu), Guangdong, Guangxi, Guizhou and Yunnan. Grows in limestone hills below alt. 500m. It likes sunshine, and it is an afforestation afforest fine species in limestone mountains for its beautiful tree-shape and red autumn leaves.

牛耳枫 *Daphniphyllum calycinum* Benth

常绿灌木。小枝灰褐色，具稀疏皮孔及叶痕。叶常集生枝顶，薄革质或厚纸质，倒卵形或倒卵状椭圆形，长11～16cm，先端近圆形或钝尖，基部宽楔形，上面绿色，下面被白粉，全缘，边缘略反卷，侧脉8～12对，网脉明显；叶柄长4～10cm。总状花序，腋生，长2～3cm；雌雄异株；花细小，无花瓣；雄花花梗长8-10mm；雌花花梗长5～6mm。核果，卵圆形，外被白粉和小疣状突起，基部具宿存的萼片。花期5～6月，果期8～11月。

产通道、城步、新宁、道县、江华、汝城、宜章。生海拔500m以下丘陵红壤灌丛、林下。分布于广东、广西、福建、江西。日本、越南亦产。株型雅致，适宜做园林观赏树种。

Evergreen shrubs. Branchlets gray brown, sparse lenticels and leaf scars. Leaves usually crowded at the top of branchlets, thinly leathery or thickly papery, obovate or obovate-elliptic, 11-16cm long, apex subrounded or obtusely pointed, base broadly cuneate, green adaxially, pruinose abaxially, margin entire, and slightly reflexed, lateral veins 8-12 pairs, reticulate veins conspicuous; petioles 4-10cm long. Racemes, axillary, 2-3cm long; dioecious; flowers small, apetalous; male pedicels 8-10mm long; female pedicels 5-6mm long. Drupe, ovate-orbicular, outside pruinose and small verrucose, base with persistent sepals. Fl. May-Jun, fr. Aug-Nov.

Distributed in Hunan (Tongdao, Chengbu, Xinning, Daoxian, Jianghua, Rucheng, Yizhang), Guangdong, Guangxi, Fujian and Jiangxi, also in Japan, Vietnam. Grows in thickets and forests in red soil below alt. 500m. It is suitable as an ornamental tree species for its elegant tree-shape.

常绿乔木。树皮灰白色，平滑。小枝幼时被褐色茸毛。叶革质，卵状椭圆形或长圆状卵形，长8～14cm，先端渐尖，基部宽楔形，全缘，上面无毛，有光泽，下面被黄褐色短柔毛，沿中脉及侧脉更密，近基部具2枚黑色腺体，侧脉5～8对；叶柄长5～8mm，被褐色茸毛。总状花序具10余花，单生或2～3簇生叶腋；花萼筒钟状，外密生黄褐色茸毛，裂片10～15；花瓣小，白色或黄色；雄蕊多数，离生。核果臀形，长8～10mm，宽10～16mm，无毛，干时深褐色。

产通道、江华、江永、桂东。生海拔520m以下低山沟谷林中。分布于福建、广东、广西、贵州、海南、云南。种子可供榨油。

Evergreen trees. Bark gray white, smooth. Branchlets brown tomentose when young. Leaves leathery, ovate-oblong or oblong-ovate, 8-14cm long, apex acuminate, base broadly cuneate, margin entire, glabrous and glossy adaxially, yellow-brown pubescent abaxially, more densely so along midvein and lateral veins, with 2 black glands nearly to base, lateral veins 5-8 pairs; petioles 5-8mm long, brown villous. Racemes, with 10 flowers, solitary or 2-3 clustered in leaf axils; calyx tube campanulate, outside densely yellow-brown tomentose, lobes 10-15; petals small, white or yellow; stamens numerous, free. Drupe reniform, 8-10mm × 10-16mm, glabrous, brown when dry.

Distributed in Hunan (Tongdao, Jianghua, Jiangyong, Guidong), Fujian, Guangdong, Guangxi, Guizhou, Hainan and Yunnan. Grows in low mountain ravine forests below alt. 520m. The seed can be used to extract oil.

常绿灌木。小枝幼时被褐色柔毛，后渐脱落。叶革质，集生枝顶，狭披针形，长3～5cm，先端圆钝或短渐尖，基部窄楔形，边缘具粗浅锯齿；叶柄扁宽，长3～4mm。顶生圆锥花序长4cm，总花梗及花梗均有棕褐色长柔毛；花萼筒状，外被褐色柔毛；花瓣椭圆披针形，长6～7mm，先端钝，白色或淡红色。果球形，黑色，径4～7mm；果梗长4～5mm，被毛。花期4～6月，果期10～11月。

产汝城、通道南部。生低山溪边林中。分布于广东、广西、海南。株型小巧，花果繁盛，可作园林观赏树种。

Evergreen shrubs. Branchlets brown pilose when young, later gradual glabrous. Leaves leathery, crowded at the top of branchlets, narrowly lanceolate, 3-5cm long, apex obtuse or short acuminate, base narrowly cuneate, margin shallowly serrate; petioles compressed and wide, 3-4mm long. Panicles terminal, 4cm long, peduncle and pedicels brown villous; calyx tubular, outside brown pilose; petals elliptic-lanceolate, 6-7mm long, apex obtuse, white or pale red. Fruit globose, black, 4-7mm in diam.; fruiting pedicels 4-5mm long, pubescent. Fl. Apr-Jun, fr. Oct-Nov.

Distributed in Hunan (Rucheng, South Tongdao), Guangdong, Guangxi and Hainan. Grows in streamside forests in low mountain. It can be used as an ornamental tree species for its small plant type and flower flourish.

深裂锈毛莓　*Rubus reflexus* Ker var. *lanceolobus* Metc.

　　落叶攀援灌木。小枝、叶下面、叶柄、花序和花萼均被锈褐色茸毛。叶革质，近圆形，长5～15cm，边缘5～7深裂，裂片披针形或长圆披针形，近等大。叶柄粗圆，长3～7cm；托叶大，长约1.4cm，掌状分裂成披针形裂片。花数朵密集生叶腋或为短总状花序；苞片与托叶同形；花梗长不及1cm。果球形，径1.5～2cm，深红色。花期7月，果期9月。

　　产江华。分布于福建、广东、广西。生海拔500m以下山地湿林下。果酸甜，可食。

　　Deciduous climbing shrubs. Branchlets, lower surface of leaves, petioles, inflorescences and calyx densely rusty brown villous. Leaves leathery, suborbicular, 5-15cm long, margin deeply 5-7-lobed, lobes lanceolate or oblong-lanceolate, subequal in length. Petioles thick and round, 3-7cm long; stipules large, ca.1.4cm long, palmately divided into lanceolate lobes. Several flowers clustered in leaf axils or short racemes; bracts the same as stipules in shape; pedicels shorter than 1cm. Fruit globose, 1.5-2cm in diam., dark red. Fl. Jul, fr. Sep.

　　Distributed in Hunan (Jianghua), Fujian, Guangdong and Guangxi. Grows in moist forests in mountain land below alt. 500m. Fruit is sweet and edible.

藤金合欢 *Acacia sinuata* (Lour.) Merr.

落叶木质藤本。枝和叶柄散生倒钩刺，被灰色短茸毛。2回羽状复叶，长15～20cm，羽片6～10对或更多；叶柄近基部及最顶端羽片之间有1腺体；小叶15～25对，线状长圆形，长8～12mm，先端圆，中脉向一侧偏斜，两面被粗毛或变无毛。头状花序，径约1cm，数个再组成腋生的大型圆锥花序；花黄色或白色；萼漏斗状，长约2mm。荚果带形，稍肉质，长7～10cm。花期6～7月，果期9～12月。

产黔阳、新宁、城步、江华、江永、临武、宜章。生海拔500m以下山地疏林中、溪边。分布于广东、广西、贵州、海南、香港、江西、云南。老挝、越南亦产。树皮含单宁。

Deciduous woody vines. Branches and petioles with scattered recurved prickles, gray velutinous. 2 pinnate, 15-20cm long, pinna 6-10 pairs or more; petioles near base and between apical pinna with 1 gland; leaflets15-25 pairs, linear-oblong, 8-12mm long, apex rounded, midvein inclined to one side, coarse hairs or glabrous on both surfaces. Capitulums, ca.1cm in diam., several capitulums arranged large panicles, axillary; flowers yellow or white; calyx funnelform, ca.2mm long. Legumes belt-shaped, slightly fleshy, 7-10cm long. Fl. Jun-Jul, fr. Sep-Dec.

Distributed in Hunan (Qianyang, Xinning, Chengbu, Jianghua, Jiangyong, Linwu, Yizhang), Guangdong, Guangxi, Guizhou, Hainan, Hongkong, Jiangxi and Yunnan, also in Laos, Vietnam. Grows in sparse forests and streamsides in mountain land below alt. 500m. Bark contains tannin.

常绿乔木。小枝无刺。芽、嫩枝、叶柄、花序均被褐色短柔毛。2回偶数羽状复叶，羽片2～4；叶柄近基部和每对羽片之间有1腺体；小叶4～10，互生或近对生，斜卵状长圆形，长1.7～10.5cm，顶生1对最大，先端渐尖且具短小尖头，基部楔形不对称，上面有光泽。头状花序单生或排列为圆锥状；花白色，无柄，萼和花冠密生锈色柔毛。荚果条形，旋卷呈环状。种子黑色，种柄丝状，悬垂于果缘。

产通道、江华、江永、郴州、汝城。生海拔500m以下山地沟谷、溪边。分布于浙江南部、福建、台湾、广东、四川、云南。印度、越南亦产。果有毒。

Evergreen trees. Branchlets unarmed. Buds, young twigs, petioles and inflorescences brown pubescent. 2 even-pinnae, pinna 2-4; petioles near base and between each pair of pinna with 1 gland; leaflets 4-10, alternate or subopposite, oblique ovate-oblong, 1.7-10.5cm long, terminal 1 pair maximum, apex acuminate and with a short cusp, base cuneate, asymmetry, glossy adaxially. Capitulums axillary or arranged paniculate; flowers white, sessile, calyx and corolla densely rusty pilose. Legumes strip-shaped, roll up like cricoid. Seeds black, seedstalks filamentous, overhanging at fruit margin.

Distributed in Hunan (Tongdao, Jianghua, Jiangyong, Chenzhou, Rucheng), South Zhejiang, Fujian, Taiwan, Guangdong, Sichuan and Yunnan, also in India, Vietnam. Grows in ravines and streamsides in mountain land below alt. 500m. The fruit is poisonous.

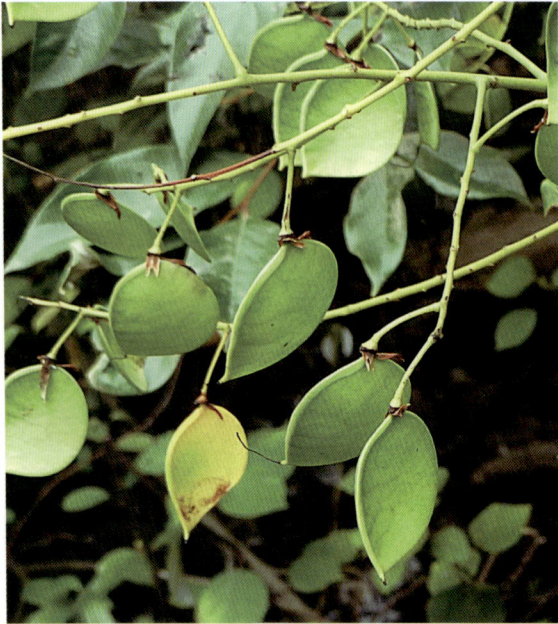

色柔毛，长10～20cm；萼片5，披针形；花瓣5，不等大，其中4片黄色，上面1片具红色斑纹；荚果革质，斜卵球形，长3～4cm，熟时棕黑色，含1种子。花期4～5月，果期7～10月。

产江华、江永、双牌、通道、宜章。生低海拔石灰岩山地。分布于广东、广西、贵州、台湾、云南。日本、东南亚亦产。

Deciduous woody vines. Branchlets sparsely recurved prickles. 2 pinnae, pinna 2-5 pairs; leaflets 4-6 pairs, elliptic or ovate, 3-6cm × 1.5-3cm, apex obtuse or mucronate, base broadly cuneate or obtuse, glabrous on both surfaces, shiny. Panicles terminal or axillary, brown pilose, 10-20cm long; sepals 5, lanceolate; petals 5, unequal in length, 4 yellow, upper 1 tinged with red stripes; Legumes leathery, obliquely ovoid, 3-4cm long, brown-black when mature, containing 1 seed. Fl. Apr-May, fr. Jul-Oct.

Distributed in Hunan (Jianghua, Jiangyong, Shuangpai, Tongdao, Yizhang), Guangdong, Guangxi, Guizhou, Taiwan and Yunnan, also in Japan, Southeast Asia. Grows in limestone mountain at low altitude.

落叶木质藤本。枝具稀疏倒钩刺。2回羽状复叶，有羽片2～5对；小叶4～6对，椭圆形或卵形，长3～6cm，宽1.5～3cm，先端钝或短尖，基部宽楔形或钝，两面无毛，有光泽。圆锥花序顶生或腋生，被褐

　　落叶木质藤本。枝、叶轴具倒钩刺和密生黄色柔毛。2回羽状复叶，有羽片8～12对；小叶15～20对，矩圆形或长椭圆形，长7～14mm，宽4～5mm，先端钝，基部偏圆形，两面疏生黄色短柔毛。顶生或腋生圆锥花序，长达30cm，花多数；花冠黄色，花瓣倒披针形。荚果革质，倒卵形，长5～6cm，被短柔毛，无刺，含1种子。花期8～9月，果期12月。

　　产宜章、江华。生低海拔路边、草丛。分布于江西南部、广东、广西、贵州。

Deciduous woody vines. Branches and leaf rachises with recurved prickles and densely yellow villous. 2 pinnae, pinna 8-12 pairs, leaflets 15-20 pairs, oblong or long elliptic, 7-14mm × 4-5mm, apex obtuse, base partial rounded, sparsely yellow pubescent on both surfaces. Panicles terminal or axillary, up to 30cm long, flowers numerous; corolla yellow, petals oblanceolate. Legumes leathery, obovate, 5-6cm long, pubescent, unarmed, containing 1 seed. Fl. Aug-Sep, fr. Dec.

Distributed in Hunan (Yizhang, Jianghua), South Jiangxi, Guangdong, Guangxi and Guizhou. Grows in roadsides, grass at low altitude.

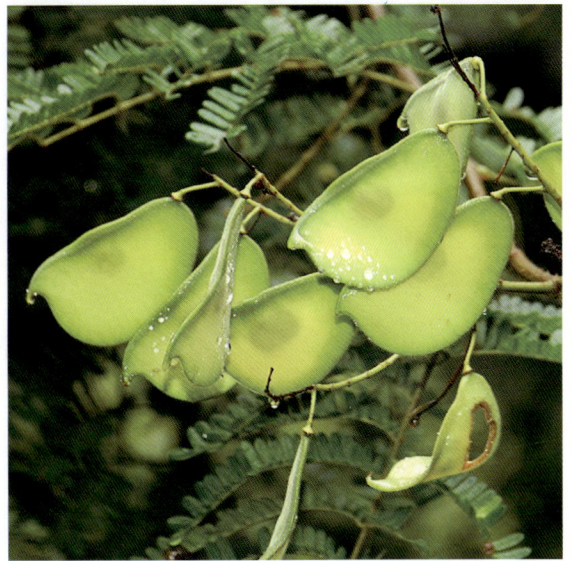

华南皂荚　　*Gleditsia fera* (Lour.) Merr.

落叶乔木。刺粗壮，分枝。1回羽状复叶，长10～20cm；小叶5～9对，纸质至薄革质，斜长椭圆形，长2～8(12)cm，先端圆钝或微凹，基部斜楔形或圆形而偏斜，边缘具圆齿，两面常无毛，有光泽，网脉细密清晰，中脉在小叶基部偏斜。花杂性，绿白色，数朵组成总状花序，长7～16cm；两性花较雄花大；子房密被棕黄色绢毛。荚果扁平，长可达45cm，劲直或稍弯，果爿革质，嫩时密被棕黄色短柔毛，老时毛渐脱落呈深棕色至黑褐色，先端具喙；种子多数，棕色至黑棕色。花期4～5月，果期6～12月。

产江永、通道、宜章。生山地林缘。分布于福建、广东、广西、香港、江西、台湾、云南。果含皂素，可制肥皂。

Deciduous trees. Spines stout, branched. 1 pinnae, 10-20cm long; leaflets 5-9 pairs, papery to thinly leathery, long elliptic, 2-8(12) cm long, apex obtuse and retuse, base obliquely cuneate or rounded and oblique, margin crenate, usually glabrous on both surfaces, shiny, reticulate veins crowded, slender and clear, midvein inclined to one side at base of blade. Flowers polygamous, green-white, several flowers consisting of racemes, 7-16cm long; bisexual flowers larger than male flowers; ovary densely yellowish-brown sericeous. Legumes compressed, up to 45cm long, strong straight or slightly curved, valves leathery, densely brown pubescent when young, glabrescent and becoming deep brown to dark brown when old, apex beaked; seeds numerous, brown to dark brown. Fl. Apr-May, fr. Jun-Dec.

Distributed in Hunan (Jiangyong, Tongdao, Yizhang), Fujian, Guangdong, Guangxi, Hongkong, Jiangxi, Taiwan and Yunnan. Grows in forest margins in mountain. The fruit contains saponin that can be used to make soap.

落叶乔木。小枝黑褐色，散生黄色皮孔。芽密被黄色疏柔毛。1回奇数羽状复叶，长25~45cm，叶轴及叶柄被黄色柔毛；小叶19~21，膜质，长圆状披针形，长6~9cm，先端渐尖或急尖，基部圆形，全缘，上面无毛，下面被灰白色糙伏毛，小叶柄长2~3mm。圆锥花序顶生；花序梗和花梗被黄色或棕色糙伏毛；萼片长圆形，被糙伏毛；花瓣红色。荚果扁平，长圆形，红棕色，长9~10cm，翅宽5~6mm，不开裂；种子圆形，平滑具光泽，棕黑色。花期5月，果期6~9月。

产城步、通道、道县、江永、江华、双牌、东安、资兴、洞口、保靖、龙山、沅陵、永顺。生低海拔山地林中、溪边、石灰岩山地。分布于广东、广西、贵州、云南。该种目前已知分布最北缘在湖南永顺小溪(北纬28.8°)。木材轻软，树干通直，为速生用材树种。国家II保护植物。

Deciduous trees. Branchlets dark brown, scattered yellow lenticels. Buds densely yellow sparse pilose. 1 odd pinnae, 25-45cm long, rachis and petioles yellow pilose; leaflets 19-21, membranous, oblong-lanceolate, 6-9cm long, apex acuminate or acute, base rounded, margin entire, glabrous adaxially, gray white strigose abaxially, petiolule 2-3mm long. Panicles terminal; peduncle and pedicels yellow or brown strigose; sepals oblong, strigose; petals red. Legumes compressed, oblong, reddish-brown, 9-10cm × 5-6mm, indehiscent; seeds round, smooth and glossy, dark brown. Fl. May, fr. Jun-Sep.

Distributed in Hunan (Chengbu, Daoxian, Tongdao, Jiangyong, Jianghua, Shuangpai, Dongan, Zixing, Dongkou, Baojing, Longshan, Yuanling, Yongshun), Guangdong, Guangxi, Guizhou and Yunnan. Grows in forests, streamsides and limestone in mountain land at low altitude. As currently known, Hunan yongshun xiaoxi (N26.6°) is the northern margin of its distribution. It is a fast-growing tree species and useful for the light wood and straight trunk. II national protected plants.

落叶小灌木。3出复叶，侧生小叶很小或缺；叶柄长1.1~2cm，上面具沟槽，疏生开展柔毛；顶生小叶长椭圆形或披针形，长5.5~10cm，先端圆形或急尖，基部钝或圆，上面无毛，下面被贴伏短柔毛，侧脉8~14对，不达叶缘，侧生小叶很小，长椭圆形。圆锥花序或总状花序顶生或腋生，花梗花时长1~4mm，花后延长至3~7mm，被开展毛；花冠紫红色。荚果镰刀形或直，长2.5~4cm，外疏被钩状短毛，具5~9种子，熟时开裂。花期7~9月，果期10~11月。

产通道、宜章。生低海拔山地灌丛、沟边。分布于福建、广东、广西、贵州、江西、四川、台湾、云南。喜马拉雅地区、东南亚亦产。侧叶在阳光下舞动，极具观赏价值。

Deciduous small shrubs. Leaves 3-foliolate, Lateral leaflets very small or absent; petioles 1.1-2cm long, grooved adaxially, sparsely pubescent, hairs spreading; terminal leaflets oblong or lanceolate, 5.5-10cm long, apex rounded or acute, base obtuse or rounded, glabrous adaxially, appressed pubescent abaxially, lateral veins 8-14 pairs, not up to leaf margin, lateral leaflets very small, long elliptic. Panicles or racemes, terminal or axillary, pedicels 1-4mm long when blooming, later extended to 3-7mm, spreading hairs; corolla purplish-red. Legumes sickle-shaped or straight, 2.5-4cm long, outside sparsely hooked pubescent, with 5-9 seeds, crack when mature. Fl. Jul-Sep, fr. Oct-Nov.

Distributed in Hunan(Tongdao, Yizhang), Fujian, Guangdong, Guangxi, Guizhou, Jiangxi, Sichuan, Taiwan and Yunnan, also in Himalaya regions, Southeast Asia. Grows in shrubs and streamsides in mountain at low altitude. It is a high-value ornamental plant, for that the side leaf can dance under the sunshine.

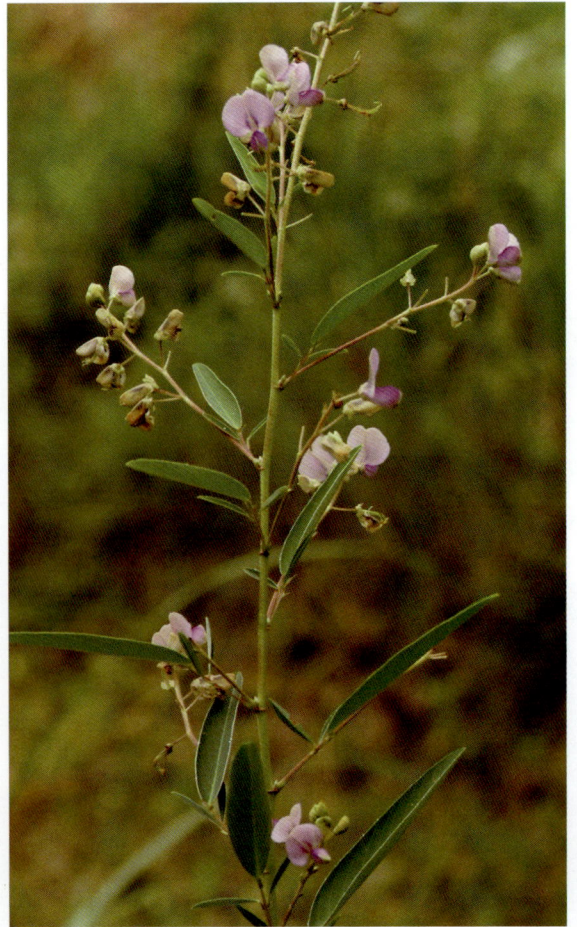

南岭黄檀　*Dalbergia balansae* Prain

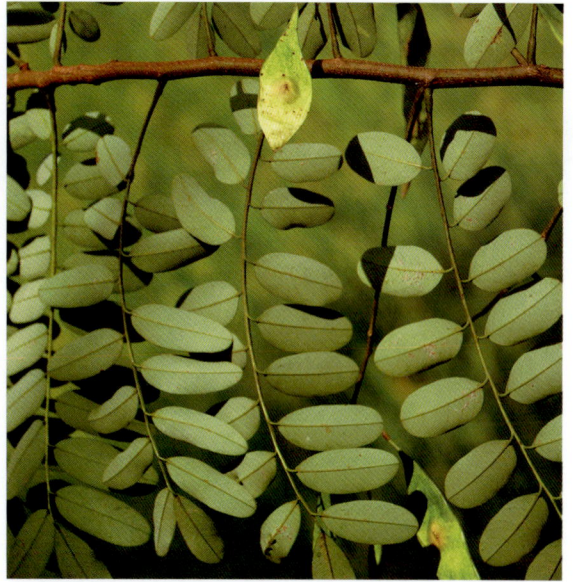

落叶乔木。1回奇数羽状复叶，长10～15cm；叶轴和叶柄被短柔毛；小叶13～15对，纸质，长圆形或倒卵状长圆形，长2～4cm，先端圆形，常微凹，基部阔楔形或圆形，老叶无毛。圆锥花序腋生，疏散，长5～10cm，总花梗和花梗疏被黄褐色短柔毛；花梗长1～2mm；花萼钟状，萼齿5；花冠白色，旗瓣圆形，顶端凹缺，翼瓣倒卵形，龙骨瓣近半月形；子房密被短柔毛。荚果带状长圆形，长5～6cm，含2～3种子，对着种子部分有明显网纹。花期6～7月，果期10～11月。

产通道、江华、江永、新宁、东安、武冈、永兴、蓝山、汝城。生海拔200～600m丘陵、山坡、路边。分布于福建、广东、广西、贵州、香港、四川、浙江南部。越南亦产。喜光，速生，为优良紫胶虫寄主树。

Deciduous trees. 1 odd pinnae, 10-15cm long; rachis and petioles pubescent; leaflets 13-15 pairs, papery, oblong or obovate-oblong, 2-4cm long, apex rounded, usually retuse, base broadly cuneate or rounded, old leaves glabrous. Panicles axillary, lax, 5-10cm long, peduncle and pedicels sparsely yellow-brown pubescent; pedicels 1-2mm long; calyx campanulate, calyx teeth 5; corolla white, upper petal orbicular, apex concave, petals obovate, keel nearly half moon; ovary densely pubescent. Legumes belt-shaped oblong, 5-6cm long, with 2-3 seeds, obvious reticulate opposite seeds. Fl. Jun-Jul, fr. Oct-Nov.

Distributed in Hunan(Tongdao, Jianghua, Jiangyong, Xinning, Dongan, Wugang, Yongxing, Lanshan, Rucheng), Fujian, Guangdong, Guangxi, Guizhou, Hongkong, Sichuan and South Zhejiang, also in Vietnam. Grows in hills, mountain slopes, roadside at alt. 200-600m. It likes sunshine, and it is a fast-growing and excellent *Kerria lacca* host.

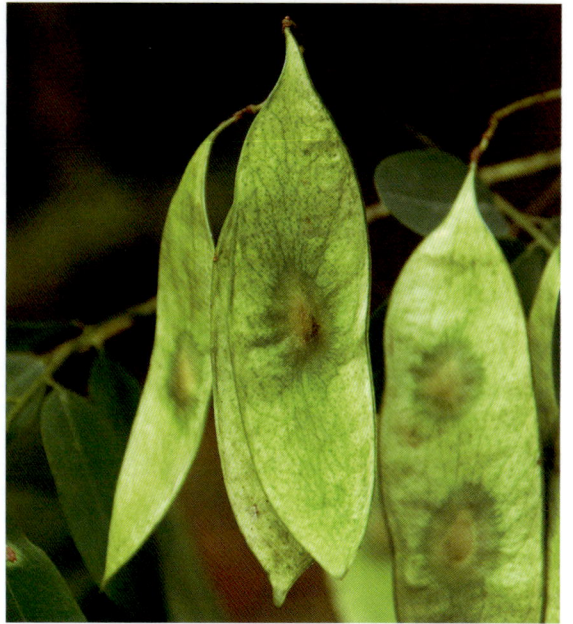

多年生直立草本。块根纺锤形，肉质；茎高20～50cm，常不分枝，密被棕色长柔毛。叶披针形，长3～7cm，先端钝或急尖，基部圆形，两面疏生短柔毛，下面沿脉密被棕色长柔毛；近无柄。总状花序腋生，极短，常具1～2花；萼钟形，萼齿5，披针形，外被棕色长柔毛；花冠淡黄色；子房密生白色长硬毛。荚果菱状椭圆形，长约8～10mm，熟时黑色，被棕色长硬毛；种子2，肾形，黑色。花期5～6月，果期7～10月。

产江华、桂东、宜章、蓝山。生低海拔山坡草丛、路边。分布于福建、广东、广西、贵州、海南、香港、江西、台湾、云南。越南、印度、澳大利亚亦产。块根可食用。

Perennial erect herbs. Root fusiform, succulent; Stem 20-50cm tall, usually unbranched, densely brown villous. Leaves lanceolate, 3-7cm long, apex obtuse or acute, base rounded, pubescent on both surfaces, densely brown villous on veins abaxially; subsessile. Racemes axillary, short, usually with 1-2 flowers; calyx campanulate, calyx teeth 5, lanceolate, outside brown villous; corolla yellow; ovary densely white hirsute. Legumes rhomboid-elliptic, ca.8-10mm long, black when mature, brown hirsute; seeds 2, reniform, black. Fl. May-Jun, fr. Jul-Oct.

Distributed in Hunan (Jianghua, Guidong, Yizhang, Lanshan), Fujian, Guangdong, Guangxi, Guizhou, Hainan, Hongkong, Jiangxi, Taiwan and Yunnan, also in Vietnam, India, Australia. Grows in grass on mountain slopes, roadsides at low altitude. Root is edible.

　　直立灌木。小枝具棱，密被紧贴柔毛。3出复叶；叶柄长3～6cm，具狭翅，被毛；顶生小叶宽披针形至椭圆形，长8～15cm，先端渐尖，基部楔形，基出3脉，两面沿脉被紧贴柔毛，下面被黑褐色小腺体，侧生小叶稍小，偏斜，基出2～3脉；小叶柄长2～5mm，密被毛。总状花序常数个聚生于叶腋，长3～8cm，常无总梗；花多而密集；花梗极短；花萼钟状；花冠紫红色。荚果椭圆形，长1～1.6cm，略被短柔毛，先端具小尖喙。花期6～9月，果期10～12月。

　　产江永、江华、通道、宜章。生海拔500m以下低丘灌丛、草丛、田边。分布于福建、广东、广西、贵州、海南、香港、江西、四川、台湾、云南。南亚、东南亚亦产。

　　Erect shrubs. Branchlets angulate, densely appressed pubescent. Leaves 3-foliolate; petioles 3-6cm long, narrowly winged, hairy; terminal leaflet broadly lanceolate to elliptic, 8-15cm long, apex acuminate, base cuneate, basal veins 3, along veins appressed pubescent on both surfaces, dark brown small glands abaxially, Lateral leaflets smaller, oblique, basal veins 2-3; petiolule 2-5mm long, densely hairy. Several racemes clustered in leaf axils, 3-8cm long, usually without peduncle; flowers numerous and dense; pedicels very short; calyx campanulate; corolla purplish-red. Legumes elliptic, 1-1.6cm long, slightly pubescent, apex with a small beak. Fl. Jun-Sep, fr. Oct-Dec.

　　Distributed in Hunan (Jiangyong, Jianghua, Tongdao, Yizhang), Fujian, Guangdong, Guangxi, Guizhou, Hainan, Hongkong, Jiangxi, Sichuan, Taiwan and Yunnan, also in South Asia, Southeast Asia. Grows in low hilly shrubs, grass, cropland edges below alt. 500m.

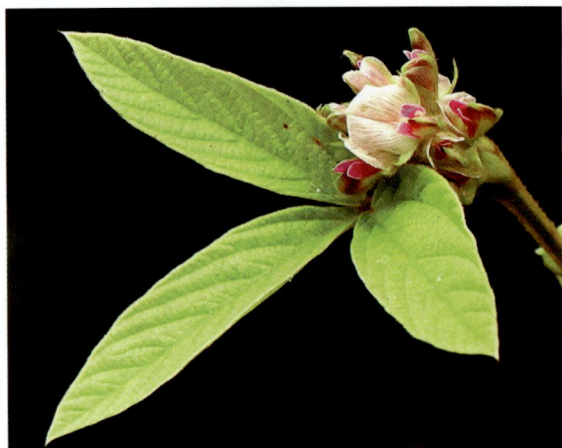

产江华、江永、宜章、桑植。生海拔500m以下低丘灌丛、草丛、田边。分布于云南、四川、贵州、湖北、广西、广东、海南、江西、福建、台湾。菲律宾亦产。

Erect or climbing shrubs. Branchlets 3-angled, densely gray brown pubescent. Leaves 3-foliolate; petioles 2-2.5cm long; terminal leaflets oblong or ovate-lanceolate, oblique, 4-8cm long, apex obtuse or abruptly mucronate, base rounded, sparsely pubescent adaxially, densely gray brown pilose abaxially, basal veins 3, and with reticulate veins slightly impressed above, Lateral leaflets smaller; petiolules very short, densely pubescent. Racemes axillary, usually 2-2.5cm long; flowers dense, with short pedicel; corolla purplish-red. Legumes elliptic, 7-8mm long, pubescent. Fl. and fr. Summer-Autumn.

Distributed in Hunan (Jianghua, Jiangyong, Yizhang, Sangzhi), Yunnan, Sichuan, Guizhou, Hubei, Guangxi, Guangdong, Hainan, Jiangxi, Fujian and Taiwan, also in Philippines. Grows in low hilly shrubs, grass, cropland edges below alt. 500m.

直立或蔓性灌木。小枝三棱形，密被灰褐色短柔毛。3出复叶；叶柄长2～2.5cm；顶生小叶长椭圆形或卵状披针形，偏斜，长4～8cm，先端钝或凸尖，基部圆形，上面被疏短柔毛，下面密被灰褐色柔毛，基出3脉，侧脉及网脉在上面多少凹陷；侧生小叶略小；小叶柄极短，密被短柔毛。总状花序腋生，常长2～2.5cm；花密生，具短梗；花冠紫红色。荚果椭圆状，长7～8mm，被短柔毛。花果期夏秋季。

乳 豆 *Galactia tenuiflora* (Klein ex Willd.) Wight et Arn.

多年生草质藤本。茎密被柔毛。3出复叶；小叶椭圆形，纸质，长2～4.5cm，两端钝圆，先端微凹，具小凸尖，上面深绿色，被疏短柔毛，下面灰绿色，密被灰白色或黄绿色长柔毛，侧脉4～7对，纤细，两面微凸；小叶柄短，长约2mm。总状花序腋生，花序轴纤细，长2～10cm，单生或孪生；花冠淡红色。荚果线形，长2～4cm，幼时被长柔毛，后渐变无毛。花果期8～11月。

产道县、江华、江永、宜章。生海拔500m以下山坡灌丛、草丛、路边。分布于江西、广东、广西、云南、台湾。东南亚、南亚亦产。

Perennial herbsaceous vines. Stem densely pilose. Leaves 3-foliolate; leaflets elliptic, papery, 2-4.5cm long, obtuse at both ends, apex retuse, with small abruptly mucronate, dark green adaxially, sparsely pubescent, greyish-green abaxially, densely gray white or yellow-green villous, lateral veins 4-7 pairs, slender, slightly raised on both surfaces; petiolules short, ca.2mm long. Racemes axillary, rachis slender, 2-10cm long, solitary or twins; corolla pale red. Legumes linear, 2-4cm long, villous when young, later gradual glabrous. Fl. and fr. Aug-Nov.

Distributed in Hunan (Daoxian, Jianghua, Jiangyong, Yizhang), Jiangxi, Guangdong, Guangxi, Yunnan and Taiwan, also in Southeast Asia, South Asia. Grows in thickets on mountain slopes, grass, roadsides below alt. 500m.

木质藤本。1回奇数羽状复叶，具3小叶，有时具5小叶；小叶近革质，阔椭圆形或椭圆形，长10～18cm，先端短突尖或钝尖，基部钝圆，两面光绿无毛，网脉明显；小叶柄长5～7mm。圆锥花序顶生，长15～30cm，花密集，单生；花梗长5～8mm；花冠淡黄色带微红或微紫色，旗瓣被绢状茸毛，阔长圆形。

荚果肿胀，椭圆形或长圆形，长4～8cm，外密被褐色细茸毛，先端具坚硬的钩状喙，具1或2～3种子，种子间缢缩。花期7～9月，果期10～12月。

产通道、江永、江华。生山地沟谷林中或林缘。分布于广东、海南、广西、贵州、云南。

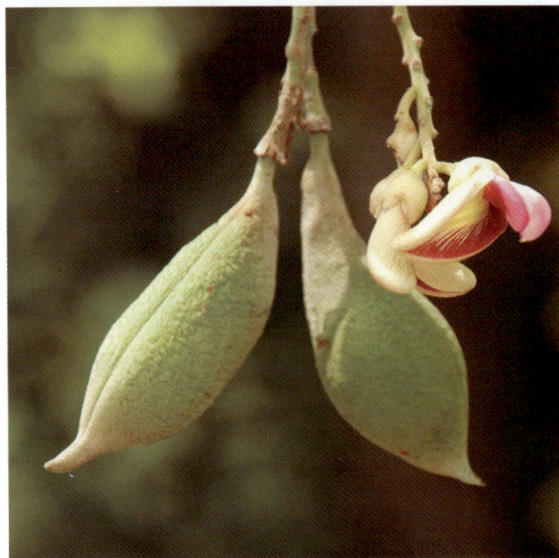

Woody vines. 1 odd pinnae, with 3 leaflets, sometimes with 5 leaflets; leaflets subleathery, broadly elliptic or elliptic, 10-18cm long, apex short abruptly acuminate or obtusely pointed, base obtuse, light green and glabrous on both surfaces, reticulate veins conspicuous; petiolules 5-7mm long. Panicles terminal, 15-30cm long, flowers dense, solitary; pedicels 5-8mm long; corolla pale yellow with reddish or purplish, upper petal silky villous, broadly oblong. Legumes swollen, elliptic or oblong, 4-8cm long, outside densely brown fine hairs, apex with hard hooked beak, with 1 or 2-3 seeds, constricted between seeds. Fl. Jul-Sep, fr. Oct-Dec.

Distributed in Hunan (Tongdao, Jiangyong, Jianghua), Guangdong, Hainan, Guangxi, Guizhou and Yunnan. Grows in ravine forests in mountain or forest margins.

常绿乔木。小枝疏生黄色柔毛。奇数羽状复叶，具3~9小叶；小叶革质，长椭圆形，长4~14cm，先端渐尖或锐尖，钝或微凹，基部楔形，无毛。圆锥花序顶生，有时腋生，花序梗和花梗密被贴伏黄色柔毛；花萼宽钟状，密生棕色柔毛；花瓣白色；子房密生黄色短柔毛。荚果小，革质，近圆形，扁平，稍膨胀，光亮，长1.5~2cm，干时暗褐色，先端具短喙；种子1，红色至红棕色，侧面压扁，径约9mm。花期4~5月，果期8~9月。

产城步、江华、通道、新宁、宜章。生山谷杂木林中。分布于福建、广东、广西、海南、香港、江西。珍贵用材树种，野生资源稀少。

Evergreen trees. Branchlets sparsely yellow pubescent. Leaves imparipinnate, with 3-9 leaflets; leaflets leathery, oblong, 4-14cm long, apex acuminate or acute, obtuse or retuse, base cuneate, glabrous. Panicles terminal, sometimes axillary, peduncle and pedicels densely appressed yellow pilose; calyx broadly campanulate, densely brown villous; petals white; ovary densely yellow pubescent. Legumes small, leathery, suborbicular, compressed, slightly swollen, shiny, 1.5-2cm long, dark brown when dry, apex shortly beaked; seed 1, red to red-brown, compressed laterally, ca.9mm in diam.. Fl. Apr-May, fr. Aug-Sep.

Distribution in Hunan (Chengbu, Jianghua, Tongdao, Xinning, Yizhang), Fujian, Guangdong, Guangxi, Hainan, Hongkong and Jiangxi. Grows in valley forests. It is a valuable timber tree, and the wild resources are scarce.

常绿乔木。幼枝、芽、幼叶、叶轴、花序轴、花梗、花萼均被黄色柔毛。1回羽状复叶，长10～24cm，小叶5～7，厚革质，矩圆形至矩圆状倒披针形，长4～12cm，下面淡白色，中脉隆起，网脉明显，叶缘略反卷。圆锥花序顶生，萼钟状，花冠紫红色，子房密生黄色柔毛。荚果木质，扁平，倒卵形或长椭圆形，长4～7cm，厚约1.5cm，密被金黄色绢质柔毛；种子1～5，红色，有光泽，长约1～1.2cm。

产通道、城步、绥宁、江永、道县、桂东、宜章、汝城、资兴。生山地沟谷、路旁或林中。分布于福建、广东、广西、贵州、海南、江西。树形高大，干直，为优良用材树种。

Evergreen trees. Young branchlets, buds, young leaves, leaf rachises, inflorescence rachis, pedicels, and calyx yellow pilose. 1 pinnae, 10-24cm long, with 5-7 leaflets, thickly leathery, oblong to oblong-oblanceolate, 4-12cm long, light white abaxially, midvein raised, reticulate veins conspicuous, margin slightly revolute. Panicles terminal, calyx campanulate, corolla purple-red, ovary densely yellow villous. Legumes woody, compressed, obovate or oblong, 4-7cm long, ca.1.5cm thick, densely golden yellow silky villous; seeds 1-5, red, glossy, ca.1-1.2cm long. Distributed in Hunan (Tongdao, Chengbu, Suining, Jiangyong, Daoxian, Guidong, Yizhang, Rucheng, Zixing), Fujian, Guangdong, Guangxi, Guizhou, Hainan and Jiangxi. Grows in ravines, roadsides or forests in mountain land. It is an excellent timber species for the tall tree-shape and straight stem.

苍叶红豆　　*Ormosia semicastrata* Hance f. *pallida* How

常绿乔木。树皮青褐色；1回羽状复叶，具7~9小叶，有时可达11小叶；小叶长椭圆状披针形或倒披针形，长4~12cm，宽1~3.5cm，基部楔形或稍钝。圆锥花序，花序各部密生黄色柔毛；花白色，子房密被黄色茸毛。荚果小，干时暗褐色，近圆形，扁平，稍膨胀，革质，长1.5~2cm，先端具短喙；种子1，红色至红棕色，侧面压扁，长约9mm。花期4~5月。

产通道、江华、江永、宜章、汝城。生海拔300~900m山地沟谷阔叶林中。分布于福建、广东、广西、海南、江西、贵州。为珍贵用材树种，野生资源稀少。

Evergreen trees. Bark green-brown; 1 pinnae, with 7-9 leaflets, sometimes up to 11 leaflets; Leaflets long elliptic-lanceolate or oblanceolate, 4-12mm × 1-3.5cm, base cuneate or slightly obtuse. Panicles, densely yellow pubescent; flowers white, ovary densely yellow villous. Legumes small, dark brown when dry, suborbicular, compressed, slightly swollen, leathery, 1.5-2cm, apex shortly beaked; Seed 1, red to red-brown, compressed laterally, ca. 9mm long. Fl. Apr-May.

Distributed in Hunan (Tongdao, Jianghua, Jiangyong, Yizhang, Rucheng), Fujian, Guangdong, Guangxi, Hainan, Jiangxi and Guizhou. Grows in broad-leaved forests in mountain ravines at alt. 300-900m. It is a valuable timber tree species, and the wild resources are scarce.

葫芦茶　*Tadehagi triquetrum*(L.) H. Ohashi

　　小灌木。茎直立，高1~2m。幼枝三棱形。单身复叶，纸质，狭披针形至卵状披针形，长5.8~13cm，宽1.1~3.5cm，通常长为宽的3倍以上，先端急尖，基部圆形或浅心形，下面沿脉被短柔毛。叶柄长1~3cm，两侧具宽翅，翅宽4~8mm。总状花序顶生和腋生，长15~30cm，被贴伏丝状毛；花2~3朵簇生于每节上；花萼宽钟形；花冠淡紫色或蓝紫色，旗瓣近圆形，先端凹。荚果长2~5cm，密被黄色或白色糙伏毛，无网脉，具5~8荚节。花期6~10月，果期10~12月。

　　产通道。生荒地或山地林缘、路旁。分布于福建、江西、广东、海南、广西、贵州、云南。印度、斯里兰卡、缅甸、泰国、越南、老挝、柬埔寨、马来西亚、太平洋岛、新喀里多尼亚岛和澳大利亚北部亦产。

Small shrubs. Stem erect, 1-2m tall. Young twigs 3-angled. Leaves 1-foliolate, papery, narrowly lanceolate to ovate-lanceolate, 5.8-13cm × 1.1- 3.5cm, usually more than 3 × as long as wide，apex acute, base rounded or shallowly cordate, pubescent on veins abaxially. Petioles 1-3cm long, with broadly wing on both sides, wing 4-8mm wide. Racemes terminal or axillary, 15-30cm long, appressed sericeous; 2-3 flowers fascicled in each node; calyx broadly campanulate; corolla pale purple or blue-purple, upper petal suborbicular, apex retuse. Legumes 2-5cm long, densely yellow or white strigose, without reticulate veins, with 5-8-jointed. Fl. Jun-Oct, fr. Oct-Dec.

　　Distributed in Hunan(Tongdao), Fujian, Jiangxi, Guangdong, Hainan, Guangxi, Guizhou and Yunnan, also in India, Sri Lanka, Burma, Thailand, Vietnam, Laos, Cambodia, Malaysia, Pacific Islands, New Caledonia and North Australia. Grows in wastelands or forest margins, roadsides in mountain land.

蕈树（阿丁枫） *Altingia chinensis* (Champ.) Oliv.

常绿乔木。小枝粗壮，无毛。叶革质，互生，狭倒卵形，长5～12cm，先端尖锐或稍钝，基部楔形，两面无毛，侧脉6～8对，边缘有浅锯齿；叶柄长约1cm。花单性同株；雄花排列成柔荑花序，无花被，雄蕊多数，花丝极短；雌花约15朵排列成头状花序，萼筒宿存，萼齿具瘤状突起；子房近下位，2室，花柱2，顶端卷曲。果序近球形，径1.7～2.5cm；种子多数，黄褐色，具光泽。花期4～6月，果期9～10月。

产通道、城步、新宁、道县、江华、江永、东安、宁远、蓝山、宜章。生山地山谷阔叶林中。分布于福建、广东、广西、贵州、海南、江西、云南、浙江南部。越南北部亦产。木材供制家具或培养香菇。

Evergreen trees. Branchlets stout, glabrous. Leaves leathery, alternate, narrowly obovate, 5-12cm long, apex acute or slightly obtuse, base cuneate, glabrous on both surfaces, lateral veins 6-8 pairs, margin shallowly dentate; petioles ca.1cm long. Flowers unisexual monoecious; male flowers arranged in ament, apetalous, stamens numerous, filaments very short; ca.15 flowers arranged in capitulum, calyx tube persistent, calyx teeth with tuberculate; ovary subinferior, 2-loculed, styles 2, apex recurved. Infructescences subglobose, 1.7-2.5cm in diam.; seeds numerous, yellowish-brown, shiny. Fl. Apr-Jun, fr. Sep-Oct.

Distributed in Hunan(Tongdao, Chengbu, Xinning, Daoxian, Jianghua, Jiangyong, Dongan, Ningyuan, Lanshan, Yizhang), Fujian, Guangdong, Guangxi, Guizhou, Hainan, Jiangxi, Yunnan and South Zhejiang, also in North Vietnam. Grows in broad-leaved forests in mountain valleys. The wood is used to make furniture or cultivate mushrooms.

大果马蹄荷　　*Exbucklandia tonkinensis* (Lecomte) Steenis

常绿乔木。小枝节膨大，具环状托叶痕。叶革质，阔卵形，长8～13cm，先端渐尖，基部阔楔形，掌状脉3～5条，全缘或幼叶掌状3浅裂，上面深绿色，下面无毛，常具小瘤状突起；叶柄长3～5cm；托叶长矩圆形，稍弯曲，长2～4cm，宽8～13mm。头状花序单生，或数个排成总状花序，具7～9花，花序柄长1～1.5cm，被黄褐色茸毛。花两性，稀单性，萼齿鳞片状；无花瓣；雄蕊约13个；子房被黄褐色柔毛。蒴果卵圆形，长1～1.5cm，外具小瘤状突起；种子6，基部具翅。花期4～5月，果期8月。

产通道、城步、道县、江华、江永、宁远、郴州、宜章。生山地阔叶林中。分布于广东、海南、广西、云南、福建、江西。老挝、越南北部亦产。树干通直，木材坚实，速生，适宜南岭地区营造用材林；叶大浓绿，托叶掌状，可作优良行道树。

Evergreen trees. Branchlets inflated on nodes, with rounded stipular scars. Leaves leathery, broadly ovate, 8-13cm long, apex acuminate, base broadly cuneate, palmately veins 3-5, margin entire or palmately 3-lobed leaves in young leaves, dark green adaxially, glabrous abaxially, usually with tuberculate; petioles 3-5cm long; stipules long oblong, slightly curved, 2-4cm × 8-13mm. Inflorescences capitate solitary, or several capitates rows in racemes, with 7-9 flowers, peduncle 1-1.5cm long, brown tomentose. Flowers bisexual, rare unisexual, calyx teeth scalelike; petals absent; stamens 13; ovary yellow-brown pubescent. Capsules ovate-orbicular, 1-1.5cm long, outside with tuberculate; seeds 6, winged at lower part. Fl. Apr-May, fr. Aug.

Distributed in Hunan(Chengbu, Daoxian, Tongdao, Jianghua, Jiangyong, Ningyuan, Chenzhou, Yizhang), Guangdong, Hainan, Guangxi, Yunnan, Fujian and Jiangxi, also in Laos, North Vietnam. Grows in broad-leaved forests in mountain land. It is suitable for building timber forests in Nanling region for its straight trunk, strong durable wood and fast-growing characteristic. It is an excellent street tree for dense, green and large leaves and palmately stipules.

　　常绿乔木。小枝无毛，皮孔明显。叶2型，革质，卵状椭圆形或掌状3裂，长8～13cm，先端渐尖，基部宽楔形，两面无毛，边缘具细锯齿，基出3脉；叶柄长2.5～4cm。花单性，雌雄同株；雄花短穗状常数个排成总状；雌花为头状花序单生，萼筒与子房合生，萼齿极短，子房2室，顶端2裂，花柱2枚。果序近球形，径约2.5cm，木质，萼齿和花柱宿存，熟时两瓣开裂。

　　产新宁、城步、绥宁、通道、江华、宜章、桂东。生山地阔叶林中。分布于福建、广东、广西、贵州、海南、江西南部。野生资源稀少，应加以保护。

Evergreen trees. Branchlets glabrous, conspicuous lenticels. Leaves dimorphic, leathery, ovate-elliptic or palmately 3-cleft, 8-13cm long, apex acuminate, base broadly cuneate, glabrous on both surfaces, margin serrulate, basal veins 3; petioles 2.5-4cm long. Flowers unisexual, monoecious; male flowers shortly spicate, usually arranged in racemes; female flowers arranged in capitates, solitary, calyx tube coadunate with ovary, calyx teeth short, ovary 2-loculed, apex 2-lobed, styles 2. Infructescences subglobose, ca.2.5cm in diam., woody, calyx teeth and style persistent, 2-valved when mature.

Distributed in Hunan(Xinning, Chengbu, Suining, Tongdao, Jianghua, Yizhang, Guidong), Fujian, Guangdong, Guangxi, Guizhou, Hainan and South Jiangxi. Grows in broad-leaved forests in mountain land. The wild resources are scarce, so it should be protected.

尖叶水丝梨　　　*Sycopsis dunnii* Hemsl.

常绿小乔木；小枝和芽被鳞毛。叶革质，卵状长圆形或卵状披针形，长6～12cm，先端锐尖或渐尖，基部宽楔形，全缘，下面幼嫩时被鳞毛，后变无毛，侧脉6～7对；叶柄长1～1.5cm。花序具苞片；雄花位于花序下半部，萼筒极短，萼齿尖锐，雄蕊约10；两性花无花瓣，具雄蕊4～8，子房密生柔毛，花柱2，无毛。蒴果卵圆形，长约1.2cm，宿存萼筒不规则裂开。

产通道、汝城、宜章。生海拔200～800m山地常绿阔叶林中。分布于福建、广东、广西、云南、贵州。

Evergreen small trees; twigs and buds scaly hairs. Leaves leathery, ovate-oblong or ovate-lanceolate, 6-12cm long, apex acute or acuminate, base broadly cuneate, margin entire, scaly hairs abaxially when young, later gradual glabrous, lateral veins 6-7 pairs; petioles 1-1.5cm long. Inflorescences with bracts; male flowers in proximal half inflorescence, calyx tube very short, calyx teeth sharp, stamens ca.10; bisexual flowers, without petals, stamens 4-8, ovary densely pilose, styles 2, glabrous. Capsule ovate-orbicular, ca.1.2cm long, persistent calyx tube irregularly fissured.

Distributed in Hunan(Tongdao, Rucheng, Yizhang), Fujian, Guangdong, Guangxi, Yunnan and Guizhou. Grows in evergreen broad-leaved forests in mountain at alt. 200-800m.

钝叶水丝梨　*Sycopsis tutcheri* Hemsl.

　　常绿灌木或小乔木；嫩枝和芽被鳞毛。叶革质，倒卵形或倒卵状椭圆形，长3～7cm，先端钝圆，基部楔形，下沿至叶柄，全缘，无毛，侧脉约5对，在两面不甚明显；叶柄长3～5mm。短总状花序长1.5～3cm；雄花生于花序下部，雄蕊3～5；两性花或雌花多生于花序上部，雄蕊4～5；子房密被柔毛；花柱2，顶端反折。蒴果卵圆形，长1～1.2cm，花柱和花萼宿存。花期4月，果期9～11月。

　　产汝城、宜章。生山地常绿阔叶林中。分布于福建、海南。

　　Evergreen shrubs or small trees; Young twigs and buds scaly hairs. Leaves leathery, obovate or obovate-elliptic, 3-7cm long, apex obtuse, base cuneate, decurrent on petiole, margin entire, glabrous, lateral veins ca.5 pairs, inconspicuous on both surfaces; petioles 3-5mm long. Short racemes, 1.5-3cm long; male flowers in lower part of inflorescence, stamens 3-5; bisexual flowers or female flowers in upper part of inflorescence, stamens 4-5; ovary densely pilose; styles 2, apex reflexed. Capsule ovoid, 1-1.2cm long, style and calyx persistent. Fl. Apr, fr. Sep-Nov.

　　Distributed in Hunan(Rucheng, Yizhang), Fujian, Hainan. Grows in evergreen broad-leaved forests in mountain.

常绿乔木。树皮灰色，不裂。叶革质，长椭圆状披针形，长8～15cm，先端长渐尖或尾尖，基部楔形至近圆形，全缘或上部具数齿，下面被黄棕色鳞秕，后变为淡棕色，侧脉9～12对；叶柄长1～2cm。雄花序穗状；雌花3朵生总苞内。壳斗近球形，径2～3cm，外被刺，刺5～8mm，基部合生成束，上部分枝，熟时不规则开裂；每壳斗含2(3)坚果。

产通道、新宁、城步、江永、江华、道县、炎陵、桂东、宜章、资兴。生低海拔山地沟谷林中。分布于福建、广东、广西、贵州、江西、台湾、云南、浙江。中南半岛亦产。喜阴湿，在南岭低山沟谷常形成纯林。

Evergreen trees. Bark gray, not crack. Leaves leathery, long elliptic-lanceolate, 8-15cm long, apex long acuminate to caudate, base cuneate to subrounded, margin entire or apically with several teeth, yellowish-brown waxy scalelike trichomes abaxially, later gradual light brown, lateral veins 9-12 pairs; petioles 1-2cm long. Male inflorescences spicate; female flowers 3 per involucre. Cupule subglobose, 2-3cm in diam., outside with spines, spines 5-8mm long, base connate into bundle, apex branched, irregularly fissured when mature, Nuts 2(3) per cupule.

Distributed in Hunan(Tongdao, Xinning, Chengbu, Jiangyong, Jianghua, Daoxian, Yanling, Guidong, Yizhang, Zixing), Fujian, Guangdong, Guangxi, Guizhou, Jiangxi, Taiwan, Yunnan and Zhejiang, also in Indo-China peninsula. Grows in ravine forests in low mountain. Liking damp, it usually can form the pure forest in low mountain of Nanling region.

常绿乔木。芽、小枝、叶柄、叶背及花序轴密被灰褐色茸毛。叶革质，长椭圆形或长圆形，长9～15cm，先端短尖至钝，基部常浅心形或圆形，全缘，侧脉8～12对；叶柄长1～4mm，粗壮。雄花序穗状或圆锥状；雌花单生总苞内。壳斗球形，径4.5～6cm，外面密被分枝的刺，刺长1～2cm，内含1坚果。

产通道、道县、宁远、江华、江永、宜章、汝城。生山沟湿林中。分布于福建、广东、广西、江西、浙江南部。果肉可生食。木材红褐色，细致坚实，属优质用材。

Evergreen trees. Buds, branchlets, petioles, leaf blades abaxially and inflorescence rachis densely gray brown villous. Leaves leathery, elliptic or oblong, 9-15cm long, apex acute to obtuse, base usually shallowly cordate or rounded, margin entire, lateral veins 8-12 pairs; petioles 1-4mm long, stout. Male inflorescences spicate or paniculate; female flower 1 per involucre. Cupule globose, 4.5-6cm in diam., outside densely branched spines, spines 1-2cm long, with 1 nut. Distributed in Hunan(Tongdao, Daoxian, Ningyuan, Jianghua, Jiangyong, Yizhang, Rucheng), Fujian, Guangdong, Guangxi, Jiangxi and South Zhejiang. Grows in wet forests in gully wet forests. The fruit can be eaten rawly. Wood is reddish brown, and meticulous solid, and belongs to the high-quality timber.

常绿乔木。树皮灰褐色，不裂。芽、小枝、幼叶均被红褐色鳞秕及短柔毛。嫩枝红紫色，具棱角。叶长椭圆形至倒卵状椭圆形，长17～25cm，先端钝尖，基部楔形，边缘具波状齿或钝锯齿，中脉粗，侧脉16～20对；叶柄长1.5～2.5cm。雌花单生总苞内。壳斗卵形至椭圆形，全包坚果，径1.2～2cm，外面被褐色鳞秕；苞片鳞片状，排成4～6条环；每壳斗含1坚果。

产通道、新宁、城步、道县、江华、江永、桂东、宜章、资兴、汝城。生海拔500m以下丘陵和低山。分布于福建、广东、广西、贵州、江西、云南、海南。越南、老挝亦产。喜光，速生，萌芽性强，耐贫瘠，系森林砍伐后萌生林的先锋树种之一。

Evergreen trees. Bark gray brown, not crack. Buds, branchlets and young leaves reddish-brown waxy scalelike trichomes and pubescent. Young twigs reddish-purple, angular. Leaves elliptic to obovate-elliptic, 17-25cm long, apex obtusely pointed, base cuneate, margin undulate-dentate or obtusely serrate, midrib stout, lateral veins 16-20 pairs; petioles 1.5-2.5cm long. Female flowers 1 per involucre. Cupule ovate to elliptic, completely wrapped nut, 1.2-2cm in diam., outside brown waxy scalelike trichomes; bracts scalelike, arranged in 4-6 rings; nut 1 per cupule.

Distributed in Hunan(Tongdao, Xinning, Chengbu, Daoxian, Jianghua, Jiangyong, Guidong, Yizhang, Zixing, Rucheng), Fujian, Guangdong, Guangxi, Guizhou, Jiangxi, Yunnan and Hainan, also in Vietnam, Laos. Grows in low hills and low mountains below alt. 500m. It is one of the pioneer trees in sprout forest for liking sun, fast growing, sprouting strongly and resistant barren.

刺栲（红锥） *Castanopsis hystrix* A. DC.

常绿乔木。树皮褐色，薄块状剥落。幼枝密被暗黄色短茸毛。叶革质，狭卵形至矩圆状披针形，长6～12cm，先端渐尖，基部圆，全缘或顶部具数浅锯齿，下面密被红棕色鳞秕和短柔毛，侧脉10～14对；叶柄长5～8mm。雄花序常为穗状。雌花单生总苞内。

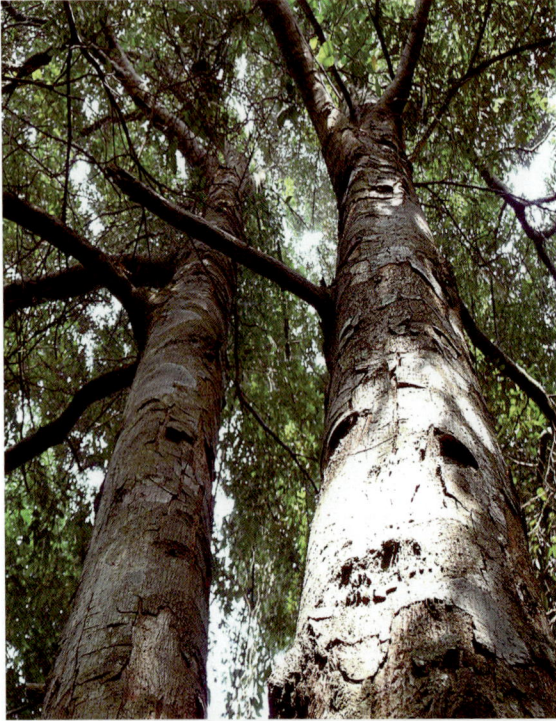

壳斗球形，径2.5～4cm，外面密被细刺，4瓣裂；内含1坚果。

产通道、新宁、江华、宜章、洞口。生海拔600m以下低山沟谷林中。分布于福建、广东、广西、贵州、海南、西藏、云南。越南至印度亦产。果实可生食。树干通直，木材红褐色、坚重，速生，系南岭及以南地区优良用材树种。

Evergreen trees. Bark brownish, thin massive spalling. Young branchlets densely dark yellow velutinous. Leaves leathery, narrowly ovate to oblong-lanceolate, 6-12cm long, apex acuminate, base rounded, margin entire or apically with few shallow serrate, densely red-brown waxy scalelike trichomes and pubescent abaxially, lateral veins 10-14 pairs; petioles 5-8mm long. Male inflorescences usually spikelike. Female flower 1 per involucre. Cupule globose, 2.5-4cm in diam., outside densely spines, 4-valved, with 1 nut.

Distributed in Hunan(Tongdao, Xinning, Jianghua, Yizhang, Dongkou), Fujian, Guangdong, Guangxi, Guizhou, Hainan, Tibet and Yunnan, also in Vietnam to India. Grows in ravine forests in low mountain below alt. 500m. The fruit can be eaten rawly. It is a high-quality timber tree in Nanling and south of Nanling region for straight trunk, fast-growing, and red-brown, strong durable wood.

常绿乔木。树皮深纵裂。叶厚革质，椭圆形至长椭圆形，长12～25cm，先端渐尖或急尖，基部楔形至圆形，全缘或上部具数齿，两面光滑无毛，侧脉11～14对；叶柄长1.5～3cm。雄花序圆锥状或穗状。雌花3朵生于总苞内。壳斗近球形，径4～6cm，外被鹿角状分枝粗刺，壳斗外壁明显可见；内含3坚果。花期3～5月，果期翌年9～11月。

产通道、城步、道县、江华、江永、宜章、资兴、炎陵、桂东、汝城。生山地沟谷湿林中。分布于福建、广东、广西、贵州南部、江西、云南。越南北部亦产。本种系南岭低山常绿阔叶林主要的优势种和建群种。果肉可生食。木材红棕色，坚重，耐腐，系优质用材。

Evergreen trees. Bark longitudinally fissured. Leaves thickly leathery, elliptic to oblong, 12-25cm long, apex acuminate or acute, base cuneate to rounded, margin entire or apically with several teeth, smooth and glabrous on both surfaces, lateral veins 11-14 pairs; petioles 1.5-3cm long. Male inflorescences paniculate or spicate. Female flowers 3 per involucre. Cupule subglobose, 4-6cm in diam., deer-hornlike stout spines, cupule clearly visible, with 3 nuts. Fl. Mar-May, fr. Sep-Nov of 2nd year.

Distributed in Hunan (Chengbu, Daoxian, Tongdao, Jianghua, Jiangyong, Yizhang, Zixing, Yanling, Guidong, Rucheng), Fujian, Guangdong, Guangxi, South Guizhou, Jiangxi and Yunnan, also in North Vietnam. Grows in ravines and moist forests in mountain land. It is a dominant tree of the evergreen broad-leaved forests in low mountain in Nanling region. Fruit can be eaten rawly. The wood is red-brown, strong durable, corrosion resistant, and belongs to the high-quality timber.

岭南青冈　　*Cyclobalanopsis championii* (Bentham) Oerst.

常绿乔木。树皮薄片状开裂。小枝密被灰褐色星状茸毛。叶片厚革质，聚生近枝顶，倒卵形或长椭圆形，长3.5～12cm，先端短钝，基部楔形，全缘反曲，稀顶端具波状齿，侧脉6～10对，中脉及侧脉在上面凹陷，下面密被星状茸毛；叶柄长0.8～1.5cm，密被褐色茸毛。壳斗碗形，包坚果1/4～1/3，径1～1.5cm，外被灰褐色短茸毛；苞片排成4～7环带。坚果宽卵形或扁球形，径1～1.5cm，基部和先端圆。花期12月至翌年3月，果期11～12月。

产汝城。生低海拔山地阔叶林中。分布于福建南部、广东、广西、海南、台湾、云南。为湖南省新记录种。

Evergreen trees. Bark thin slice fissured. Branchlets densely gray stellate tomentose. Leaves thickly leathery, crowded toward branchlet apex, obovate or oblong, 3.5-12cm long, apex obtuse, base cuneate, margin recurved and entire or rarely undulate-crenate toward apex, lateral veins 6-10 pairs, midvein and lateral veins impressed adaxially, densely stellate tomentose abaxially; petioles 0.8-1.5cm long, densely brownish tomentose. Cupule bowl-shaped, enclosing 1/4-1/3 of nut, 1-1.5cm in diam., outside brownish velutinous; bracts arranged in 4-7 rings. Nut broadly ovate or depressed globose, 1-1.5cm in diam., base and apex rounded. Fl. Dec-Mar of 2nd year, fr. Nov-Dec.

Distributed in Hunan(Rucheng), South Fujian, Guangdong, Guangxi, Hainan, Taiwan and Yunnan. Grows in broad-leaved forests in mountain land at low altitude. A new recorded species in Hunan province.

常绿乔木。芽、幼枝、幼叶及柄、花序均密被黄色茸毛，易脱落。叶薄革质，长椭圆形或披针形，长7～13cm，先端钝尖，基部楔形，全缘或顶部具数浅齿，略反卷，侧脉6～10对，下面初被黄色茸毛，后渐脱落，叶柄长1～1.4cm。雄花序2～4个簇生；雌花序长1～2cm，具2～5花。果序长1～1.5cm，具1～2果。壳斗浅碗形至盘形，包坚果基部，密被黄褐色茸毛，具4～5环带。坚果扁球形，径1.5～3cm，幼时密被黄褐色茸毛，后渐无毛。花期4～5月，果期9～12月。

产江华、汝城、新宁、资兴。生海拔200～500m山地沟谷林缘。分布于广东、广西。壳斗及果形奇雅致，可供陈列观赏。

Evergreen trees. Buds, young branchlets, young leaves, petioles, and inflorescences densely yellow tomentose, easy to fall off. Leaves thinly leathery, oblong or lanceolate, 7-13cm long, apex obtusely pointed, base cuneate, entire or apically with several shallowly teeth, slightly recurved, lateral veins 6-10 pairs, lower surface yellow villous at first, later gradual glabrous; petioles 1-1.4cm long. Male inflorescences 2-4-fascicled; female inflorescences 1-2cm long, with 2-5 flowers. Infructescences 1-1.5cm long, with 1-2 fruits. Cupule shallowly bowl-shaped to disciform, enclosing the base of nut, densely yellow-brown tomentose, with 4-5 rings. Nut oblate, 1.5-3cm in diam., densely yellow-brown tomentose when young, later glabrescent. Fl. Apr-May, fr. Sep-Dec.

Distributed in Hunan(Jianghua, Rucheng, Xinning, Zixing), Guangdong and Guangxi. Grows in ravines and forest margins in mountain land at alt. 200-500m. The cupule and fruit are ornamental for its strange shapes.

饭甑青冈　*Cyclobalanopsis fleuryi* (Hickel et A. Camus) Chun

常绿乔木。树皮灰白色，平滑。芽、小枝、幼叶、花序、壳斗密被棕色长茸毛。叶革质，长椭圆形或卵状长椭圆形，长15～25cm，先端急尖或短渐尖，基部楔形，全缘或顶端具波状细齿，初密被黄棕色茸毛，后脱落，叶背粉白色，侧脉9～12对；叶柄长2～6cm。雄花序长10～15cm；雌花序长2.5～3.5cm，着生4～5花。壳斗饭钵形，包着坚果约2/3，高3～4cm，具8～10环带。坚果圆柱形，密被黄棕色茸毛。花期3～4月，果期10～11月。

产道县、江华、江永、汝城。生海拔500m以下山地沟谷林中。分布于福建、广东、广西、贵州、海南、江西、云南。越南亦产。本种喜湿热气候，湖南南岭地区局部"暖窝子"处才能生长。

Evergreen trees. Bark gray white, smooth. Buds, twigs, young leaves, inflorescence, cupule densely brown long tomentose. Leaves leathery, oblong or ovate-oblong, 15-25cm long, apex acute or short acuminate, base cuneate, margin entire or apically undulate and serrulate, densely orangish brown tomentose when young but glabrescent, whitish abaxially, lateral veins 9-12 pairs; petioles 2-6cm long. Male inflorescences 10-15cm long; female inflorescences 2.5-3.5cm long, with 4-5 flowers. Cupule bowl-shaped, enclosing ca.2/3 of nut, 3-4cm tall, with 8-10 rings. Nut terete, 3-4.5cm tall, densely yellow-brown tomentose. Fl. Mar-Apr, fr. Oct-Nov.

Distributed in Hunan(Daoxian, Jianghua, Jiangyong, Rucheng), Fujian, Guangdong, Guangxi, Guizhou, Hainan, Jiangxi and Yunnan, also in Vietnam. Grows in ravine forests in mountain land below alt. 500m. Because of liking hot and humid climate, it can only grow well in "local warm habitat" of Hunan Nanling local region.

烟斗石栎 *Lithocarpus corneus* (Lour.) Rehd.

常绿乔木。叶革质，长椭圆形至倒卵状披针形，长6～18cm，先端渐尖，基部宽楔形，边缘具锯齿或浅波状，老叶无毛，侧脉9～15对，连中脉在上面下凹；叶柄长8～10mm。雄花序穗状，下部常3雌花簇生。壳斗陀螺形，约半包坚果，径2.5～3.5cm；苞片三角形至菱形；坚果近球形或陀螺形，顶端圆，扁平，或稍凹，被微柔毛；果期凸起。花期5～7月，果期翌年9～10月。

产通道、道县、江华、江永。生海拔700m以下低山沟谷林中。分布于福建、广东、广西、贵州、海南、台湾、云南。越南、老挝亦产。本种为典型华南物种，湖南南岭地区位于其分布区的北缘，只有在局部热湿小气候条件下才能生长，指示当地平均气温高于18℃，1月份平均温度高于8℃。

Evergreen trees. Leaves leathery, oblong to obovate-lanceolate, 6-18cm long, apex acuminate, base broadly cuneate, margin dentate or shallowly undulate, old leaves glabrous, lateral veins 9-15 pairs, and with midvein raised adaxially; petioles 8-10mm long. Male inflorescences spicate, often 3 female flowers tufted at lower part. Cupule turbinate, enclosing ca.1/2 of nut, 2.5-3.5cm in diam.; bracts triangular to rhomboid; Nut subglobose to turbinate, apex rounded, compressed, or slightly concave, puberulent; scar convex. Fl. May-Jul, fr. Sep-Oct of 2nd year.

Distributed in Hunan(Tongdao, Daoxian, Jianghua, Jiangyong), Fujian, Guangdong, Guangxi, Guizhou, Hainan, Taiwan and Yunnan, also in Vietnam, Laos. Grows in ravine forests in low mountain below alt. 700m. This is a typical species of Southern China, and Hunan Nanling region is located in the northern margin of its distribution area, and only grow well in the local thermal microclimate conditions. Indicate the local average temperature higher than 18℃, the average temperature of January higher than 8℃.

常绿乔木。小枝、幼叶、叶柄、叶背及花序轴均密被黄褐色长柔毛。叶硬革质，椭圆形、卵形或倒卵状椭圆形，长8～15cm，先端圆或短突尖，基部圆或阔楔形，叶缘背卷，侧脉9～13对，连细脉在上面凹陷，在下面突起，叶片呈泡状隆起；叶柄长2～3cm。幼嫩壳斗全包幼小的坚果，成熟壳斗盆状，高3～6mm，壁近木质，苞片线状反折，顶端尖钩状。坚果扁球形，径2～3cm。花期7～9月，果次年同期成熟。

产通道、江华、宜章。生海拔500m以下山地及河滩林中。分布于广东、广西、贵州。湖南南岭地区少见。

Evergreen trees. Branchlets, young leaves, petioles, leaf blades abaxially, and inflorescence rachis densely yellow-brown villous. Leaves stiffly leathery, elliptic, ovate or obovate-elliptic, 8-15cm long, apex rounded or short abruptly acuminate, base rounded or broadly cuneate, margin recurved, lateral veins 9-13 pairs, and with small veins impressed above, raised below, blades bubble-like bulge; petioles 2-3cm long. Young cupules completely enclosing young nut, mature cupules bowl-shaped, 3-6mm tall, wall nearly woody, bracts linear reflexed, tip hooked. Nut oblate, 2-3cm in diam. Fl. Jul-Sep, fr. Jul-Sep of following year.

Distributed in Hunan(Tongdao, Jianghua, Yizhang), Guangdong, Guangxi and Guizhou. Grows in mountain land and river beach in the forests below alt. 500m. It is a rare species in Nanling region of Hunan.

常绿乔木。小枝、幼叶、叶柄、叶背及花序轴被黄褐色长柔毛。叶硬纸质，狭长椭圆形或披针形，长7～16cm，先端长渐尖，基部楔形，全缘，侧脉11～13对，下面密被平伏柔毛及蜡鳞；叶柄长1～1.5cm。壳斗近球形或呈梨形，径2.5～3mm，全包坚果或包坚果3/4，壁薄，苞片三角形，伏贴，被鳞秕，位于上部的小苞片具锥状尖头，长约3mm；坚果扁球形，径2～2.5cm，栗褐色。花期8～9月，果次年10～11月。

产通道、新宁、绥宁、宜章。生海拔200～800m山地阔叶林中。分布于福建、广东、广西、江西、贵州、四川南部。

Evergreen trees. Branchlets, young leaves, petioles, lower surface of leaves, and inflorescence rachis yellow-brown villous. Leaves stiffly papery, narrowly elliptic or lanceolate, 7-16cm long, apex long acuminate, base cuneate, margin entire, lateral veins 11-13 pairs, appressed pubescent and waxy scales abaxially; petioles 1-1.5cm long. Cupule subglobose or pear-shaped, 2.5-3mm in diam., completely enclosing nut or enclosing 3/4 of nut, wall thin, bracts triangular, appressed, with scales, upper bracteoles conical cusp, ca.3mm long; Nut oblate, 2-2.5cm in diam., maroon brown. Fl. Aug-Sep, fr. Oct-Nov of 2nd year.

Distributed in Hunan(Tongdao, Xinning, Suining, Yizhang), Fujian, Guangdong, Guangxi, Jiangxi, Guizhou and South Sichuan. Grows in broad-leaved forests in mountain land at alt. 200-800m.

滑皮石栎　*Lithocarpus skanianus* (Dunn) Rehd.

常绿乔木。小枝、芽、幼叶、叶柄、花序轴均密被黄棕色茸毛。叶厚纸质，倒卵状椭圆形或倒披针形，长6～16cm，先端短尾状或渐尖，基部楔形，全缘，侧脉10～13对，在上面微凹，小脉近平行，下面密被灰白色鳞秕，沿脉被棕色长柔毛；叶柄6～9mm。壳斗杯状至近球形，近全包坚果，苞片钻形或锥形，贴于壳壁，被短毛及鳞秕；坚果扁球形至宽圆锥形；果脐凸。花期9～10月，果翌年同期成熟。

产通道、江华、江永、宜章、桂东、汝城。生海拔800m以下山地阔叶林中。分布于福建南部、广东、广西、海南、江西南部、云南东南部。

Evergreen trees. Branchlets, buds, young leaves, petioles, and inflorescence rachis densely tawny tomentose. Leaves thickly papery, obovate-elliptic or oblanceolate, 6-16cm long, apex shortly caudate or acuminate, base cuneate, margin entire, lateral veins 10-13 pairs, impressed above, small veins subparallel, densely gray-white scurfy and brown villous on veins below; petioles 6-9mm long. Cupule cupular to subglobose, almost completely enclosing nut, bracteoles subulate or tapered, appressed, with puberulent and scurfy; Nut depressed globose to broadly conical, scar convex. Fl. Sep-Oct, fr. Sep-Oct of following year.

Distributed in Hunan(Tongdao, Jianghua, Jiangyong, Yizhang, Guidong, Rucheng), South Fujian, Guangdong, Guangxi, Hainan, South Jiangxi and Southeast Yunnan. Grows in broad-leaved forests in mountain land below alt. 800m.

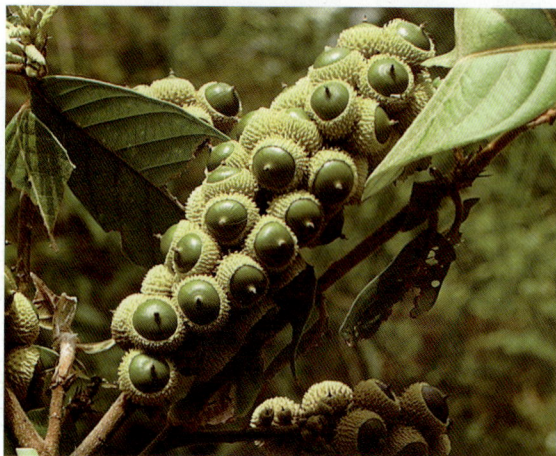

紫玉盘石栎（紫玉盘柯） *Lithocarpus uvariifolius* (Hance) Rehd.

常绿乔木。小枝、芽、叶柄及花序轴均密被棕黄色或锈色长毛。叶革质或厚纸质，椭圆形至长椭圆形，或倒卵形，长12～22cm，先端突尖，基部近圆，全缘或顶部具浅钝齿，侧脉密集，25～28对，连中脉在上面凹陷，下面密生灰黄色茸毛；叶柄长1.5～3cm。果序有成熟壳斗1～4；壳斗深碗状或半圆形，包坚果1/2以上；苞片大，幼时狭椭圆形到披针形，老时呈菱状隆起；坚果半球形，被黄色细茸毛。

产汝城。生于低海拔山地常绿阔叶林中。分布于广东、广西、福建西南部。湖南南岭地区稀见种。湖南省新记录种。

Evergreen trees. Branchlets, buds, petioles, and inflorescence rachis densely brown-yellow or rust-colored long hairs. Leaves leathery or thickly papery, elliptic to oblong, or obovate, 12-22cm long, apex abruptly acuminate, base suborbicular, margin entire or apically shallowly crenatum, lateral veins dense, 25-28 pairs, and with midvein impressed above, densely gray yellow tomentose below; petioles 1.5-3cm long. Infructescence with 1-4 developed cupules; deep bowl-shaped or semirounded, enclosing more than 1/2 of nut; bracts large, narrowly oblong to lanceolate when young, and diamond-shaped bulge when old; Nut hemiglobose, densely yellow fine hair.

Distributed in Hunan(Rucheng), Guangdong, Guangxi and Southwest Fujian. Grows in evergreen broad-leaved forests in low mountain. It is a rare species in Nanling region of Hunan. A new recorded species in Hunan province.

二色波罗蜜　*Artocarpus styracifolius* Pierre

常绿乔木。小枝黑褐色，无毛。叶互生，排为2列，长圆形或卵状披针形，长4～8cm，先端尾状渐尖，基部楔形下延于叶柄，全缘，上面近无毛，下面密被灰白色粉末状毛，脉上更密，侧脉4～7对，网脉明显；叶柄长1～1.4cm，被毛。花雌雄同株；雄花序椭圆形，长6～12mm，密被灰白色短柔毛；总花梗长6～12mm。聚花果球形，径约4cm，黄色，干时红褐色，密被毛，外密生弯曲圆形突起；果梗长2～2.5cm，上部增粗，被柔毛。花期夏季，果期秋季。

产通道、洞口、江华。生低海拔常绿林中。分布于广东、广西、海南、云南。中南半岛亦产。果酸甜可食。湖南南岭地区稀见。目前已知分布最北界为洞口（北纬27.06°）。

Evergreen trees. Branchlets dark brown, glabrous. Leaves alternate, distichous, oblong or ovate-lanceolate, 4-8cm long, apex caudate-acuminate, base cuneate to decurrent on petiole, margin entire, subglabrous adaxially, densely white farinaceous hairs abaxially, veins more dense, lateral veins 4-7 pairs, reticulate veins conspicuous; petioles 1-1.4cm, pubescent. Flowers monoecious; male inflorescences elliptic, 6-12mm long, densely gray pubescent; peduncle 6-12mm long. Syncarp globose, ca.4cm in diam., yellow, red-brown when dry, densely hairy, outside densely curved round protrusions; carpopodium 2-2.5cm long, upper part widening, pubescent. Fl. Summer, fr. Autumn. Distributed in Hunan(Tongdao, Dongkou, Jianghua), Guangdong, Guangxi, Hainan and Yunnan, also in Indo-China Peninsula. Grows in evergreen forests at low altitude. Fruit is sweet and edible. It is a rare species in Nanling region of Hunan. The north boundary of its distribution is located in Dongkou (N 27.06°).

常绿乔木。树皮深紫色，片状剥落。幼枝被柔毛。叶革质，椭圆形至倒卵形，长8～15cm，先端渐尖至短渐尖，基部楔形，全缘，幼树之叶常为羽状浅裂，下面被灰色粉末状柔毛，侧脉6～7对，弯拱向上，与网脉在下面明显突起；叶柄长1.5～2cm，被毛。花序单生叶腋；雄花序椭圆形，长1.5～2cm，总梗长2～4.5cm，被短柔毛。聚花果近球形，径3～4cm，浅黄色至金黄色，外被棕色短柔毛和乳突；果柄长3～5cm，被短柔毛。花期4～5月，果期8～10月。

产江华、江永、宜章、资兴、永兴、炎陵、汝城、桂东。生低山阔叶林中或石灰岩山地。分布于广东、海南、福建、江西、云南。本种湖南南岭地区野生资源稀少，应加强保护。

Evergreen trees. Bark dark purple, flake off. Young branchlets pilose. Leaves leathery, elliptic to obovate, 8-15cm long, apex acuminate to short acuminate, base cuneate, margin entire, pinnately lobed on young trees, gray farinaceous pilose adaxially, lateral veins 6-7 pairs, arching upward, and with reticulate veins obvious raised below; petioles 1.5-2cm long, pubescent. Panicles solitary and axillary; male inflorescences elliptic, 1.5-2cm long, peduncle 2-4.5cm long, pubescent. Syncarp globose, 3-4cm in diam., pale yellow to golden yellow, brown pubescent and papillate; carpopodium 3-5cm, pubescent. Fl. Apr-May, fr. Aug-Oct.

Distributed in Hunan(Jianghua, Jiangyong, Yizhang, Zixing, Yongxing, Yanling, Rucheng, Guidong), Guangdong, Hainan, Fujian, Jiangxi and Yunnan. Grows in broad-leaved forests in low mountain or in limestone mountains. The wild resources are scarce in Nanling region of Hunan, so it should be protected.

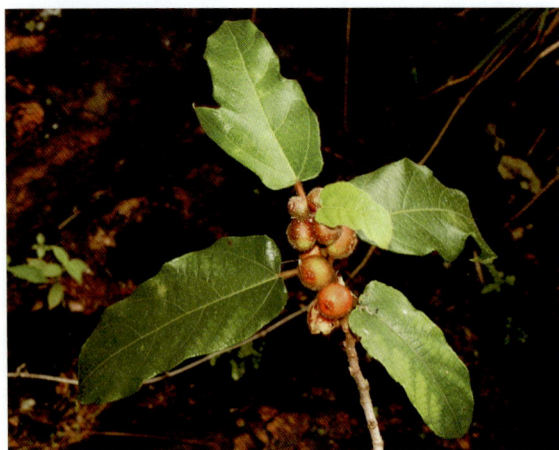

常绿灌木。小枝、叶、叶柄和榕果均被金黄色粗毛。叶纸质，卵形、椭圆形或卵状椭圆形，长10~25cm，先端急尖或渐尖，基部圆形、浅心形或宽楔形，全缘或3~5裂，基生3~5脉，边缘具细锯齿，无毛或金黄色长硬毛；叶柄长2~8cm。榕果成对腋生枝上，球形或椭圆球形，无梗或近无梗，径1~1.5cm，密被黄色长硬毛，幼时顶部脐状。

产通道、城步、道县、江华、江永、汝城、资兴。生海拔800m以下山地林缘、次生林或沟边。分布于云南、贵州、广西、广东、海南、福建、江西。喜马拉雅地区、东南亚亦产。茎皮纤维制麻绳、麻袋。

Evergreen shrubs. Branchlets, leaves, petioles and figs golden yellow hirsute. Leaves papery, ovate, elliptic or ovate-elliptic, 10-25cm long, apex acute or acuminate, base rounded, shallowly cordate or broadly cuneate, margin entire or 3-5-lobed, basal veins 3-5, margin serrulate, glabrous or golden yellow hirsute; petioles 2-8cm long. Figs axillary on normal leafy shoots, paired, globose or elliptic, sessile or subsessile, 1-1.5cm in diam., densely yellow hirsute, apical pore navel-like when young.

Distributed in Hunan(Tongdao, Chengbu, Daoxian, Jianghua, Jiangyong, Rucheng, Zixing), Yunnan, Guizhou, Guangxi, Guangdong, Hainan, Fujian and Jiangxi, also in Himalaya region, Southeast Asia. Grows in forest margins, secondary forest, and streamsides in mountain below alt. 800m. Stem skin fiber can manufacture hemp rope and sacks.

常绿乔木。小枝黄褐色，被红褐色糠屑状毛。叶纸质，互生，椭圆状披针形至椭圆形，长7～19cm，先端尾状渐尖，基部宽楔形，全缘，两面无毛，干时浅褐色至深褐色，基出3脉，达叶的近1/2，侧脉2～4对，与网脉在下面明显突起；叶柄长1～4cm，无毛。榕果成对或单生叶腋，球形，顶端脐状凸起，径5～10mm，被锈色糠屑状毛，基生苞片3，总梗较细，长5～15mm，被锈色糠屑状毛。

产通道、城步、江永。生海拔500m以下石山或山谷溪水边。分布于福建、广东、广西、海南、四川、云南。印度、老挝、越南亦产。

Evergreen trees. Branchlets yellow-brown, red-yellow scurfy hairs. Leaves papery, alternate, elliptic-lanceolate to elliptic, 7-19cm long, apex caudate-acuminate, base broadly cuneate, margin entire, glabrous on both surfaces, light brown to dark brown when dry, basal veins 3, extending nearly 1/2 length of leaf blade, lateral veins 2-4 pairs, and with reticulate veins abaxially conspicuous; petioles 1-4cm long, glabrous. Figs axillary on normal leafy shoots, paired or solitary, globose, apical pore navel-like, convex, 5-10mm in diam., with rust-colored scurfy hairs, basal bracts 3, peduncle slender, 5-15mm long, with rust-colored scurfy hairs. Distributed in Hunan(Tongdao, Chengbu, Jiangyong), Fujian, Guangxi, Hainan, Guangdong, Sichuan and Yunnan, also in India, Laos, Vietnam. Grows in rocky mountains or valley streamsides below alt. 500m.

常绿灌木。小枝被糙毛。叶纸质，互生，倒披针形或倒卵状披针形，长4～11cm，先端突尖或短尾状渐尖，基部楔形至圆形，全缘，下面被微柔毛和小钟乳体，基出3脉，侧脉5～9对，与网脉在下面不明显；叶柄长3～5mm，被毛。榕果单生于叶腋，梨形，径约1.5cm，果柄长8～15mm，被柔毛，基部苞片3。

产通道、江华、新宁、宜章、汝城。生低海拔溪边、潮湿林下。分布于福建、广东、广西、海南、云南、贵州。越南、马来半岛、印度亦产。

Evergreen shrubs. Branchlets coarsely hairy. Leaves papery, alternate, oblanceolate or obovate-lanceolate, 4-11cm long, apex abruptly acuminate or shortly caudate-acuminate, base cuneate to rounded, margin entire, puberulent and small cystoliths abaxially, basal veins 3, lateral veins 5-9 pairs, and with reticulate veins inconspicuous below; petioles 3-5mm long, hair. Figs axillary on normal leafy shoots, solitary, pear-shaped, ca.1.5cm in diam., peduncle 8-15mm long, pubescent, basal bracts 3.

Distributed in Hunan(Tongdao, Jianghua, Xinning, Yizhang, Rucheng), Fujian, Guangdong, Guangxi, Hainan, Yunnan and Guizhou, also in Vietnam, Malay peninsula, India. Grows in streamsides or under moisture forests at low altitude.

　　落叶乔木。幼时附生，具板状或支柱根。叶薄革质或厚纸质，卵状披针形至椭圆状卵形，长8～16cm，先端短渐尖，基部钝圆、楔形至浅心形，全缘，两面无毛，基出3脉，较短，侧脉7～10对，与网脉在两面稍明显；叶柄长 2～5cm，无毛。榕果单生叶腋或簇生于无叶腋部，近球形，径5～12mm，熟时紫红色，基生苞片3，细小；无总梗或有短梗。花期5～8月。

　　产通道、江永。生海拔200～400m溪边或石灰岩山地。分布于福建、广东、广西、云南、海南、台湾、浙江。东南亚至大洋洲亦产。树冠宽大浓密，常用作行道树。湖南南岭地区少见种。

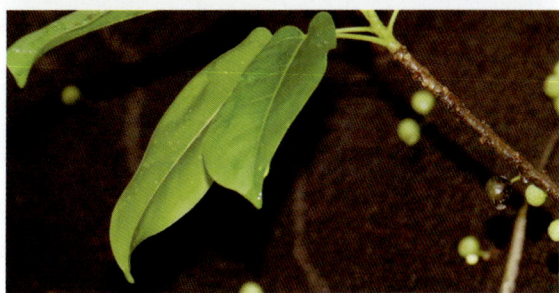

Deciduous trees. Epiphytic when young, with buttress or prop roots. Leaves thinly leathery or thickly papery, ovate-lanceolate to elliptic-ovate, 8-16cm long, apex shortly acuminate, base obtusely rounded, cuneate, or shallowly cordate, margin entire, glabrous on both surfaces, basal veins 3, short, lateral veins 7-10 pairs, and with reticulate veins slightly conspicuous on both surfaces; petioles 2-5cm long, glabrous. Figs axillary on leafy branchlets, solitary or in clusters on leafless older branchlets, subglobose, 5-12mm in diam., red to purple when mature, basal bracts 3, small; sessile or short pedunculate. Fl. May-Aug.

Distributed in Hunan(Tongdao, Jiangyong), Fujian, Guangdong, Guangxi, Yunnan, Hainan, Taiwan and Zhejiang, also in Southeast Asia to Australia. Grows in streamsides or limestone mountains at alt. 200-400m. It is usually used as street ornamental trees for its wide, dense, foliage crown. It is a rare species in Nanling region of Hunan.

秤星树（梅叶冬青） *Ilex asprella* (Hook. et Arn.) Champ. ex Benth.

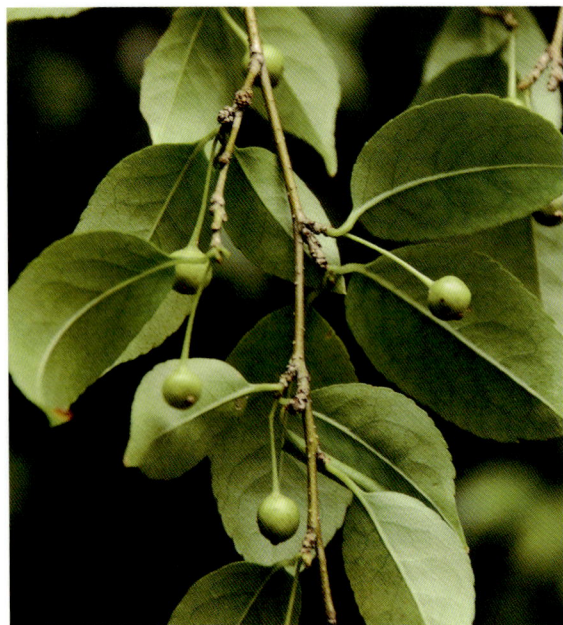

落叶灌木。小枝纤细，栗褐色，皮孔明显。叶膜质，卵形或卵状椭圆形，长3～7cm，先端尾状渐尖，基部钝，边缘具锯齿，中脉在上面微凹且被毛，侧脉5～6对，于近叶缘处网结，网状脉两面可见；叶柄长3～8mm，上面有沟。雄花序2～3花簇生，花梗长4～7mm，花冠白色；雌花序单生于叶腋，花梗长1～2cm，花冠辐状，花瓣近圆形。果球形，径5～7mm，熟时黑色，具4～6纵条纹及沟，宿存柱头头状或厚盘状，分核4～6粒。花期3月，果期4～10月。

产通道、江永、永兴、郴州。生海拔500m以下丘陵草丛、灌丛。分布于浙江、江西、福建、台湾、广东、广西、香港。菲律宾亦产。

Deciduous shrubs. Branchlets slender, maroon brown, lenticels conspicuous. Leaves membranous, ovate or oblong-ovate, 3-7cm long, apex caudate-acuminate, base obtuse, margin serrate, midvein impressed and with hairy adaxially, lateral veins 5-6 pairs, anastomosing near margin, reticulate veins visible on both surfaces; petioles 3-8mm long, grooved adaxially. Male inflorescences 2-3-fascicled, pedicels 4-7mm long, corolla white; female inflorescences solitary in leaf axils, pedicels 1-2cm long, corolla rotate-campanulate, petals suborbicular. Fruit globose, 5-7mm in diam., black when mature, with 4-6 longitudinal stripes and sulcate, persistent stigma capitate or thickly disciform, style evident, pyrenes 4-6. Fl. Mar, fr. Apr-Oct.

Distributed in Hunan(Tongdao, Jiangyong, Yongxing, Chenzhou), Zhejiang, Jiangxi, Fujian, Taiwan, Guangdong, Guangxi and Hongkong, also in Philippines. Grows in hilly grass or shrubs below alt. 500m.

果球形，径5～6mm，熟时红色，平滑，有光泽，宿存柱头厚盘状，分核4～5。花期5月，果期8～12月。

产新宁、江华。生海拔500m以下山地林下、林缘及阴湿处。分布于福建、江西、广东、广西。

常绿小乔木。小枝、芽、叶柄、叶片、花梗和花萼均密被黄锈色茸毛。叶薄革质，卵形、卵状椭圆形或卵状披针形，长3～9cm，先端渐尖或急尖，基部钝或楔形，全缘具锈色缘毛，两面侧脉不明显；叶柄长3～5mm。雌雄异株；聚伞花序单生于当年生枝叶腋。

Evergreen small trees. Branchlets, buds, petioles, leaf blades, pedicels, and calyx densely yellow rusty tomentose. Leaves thinly leathery, ovate, ovate-elliptic or ovate-lanceolate, 3-9cm long, apex acuminate or acute, base obtuse or cuneate, margin entire and with rust-colored ciliate, lateral veins inconspicuous on both surfaces; petioles 3-5 mm long. Dioecious; cymes, solitary, axillary on current year's branchlets. Fruit globose, 5-6mm in diam., red when mature, smooth, glossy, persistent stigma thickly disciform, pyrenes 4-5. Fl. May, fr. Aug-Dec.

Distributed in Hunan(Xinning, Jianghua), Fujian, Jiangxi, Guangdong and Guangxi. Grows in forests, forest margins and moist places in mountain land below alt. 500m.

毛冬青　*Ilex pubescens* Hook. et Arn.

常绿灌木。小枝细长，近4棱，密被短硬毛。叶膜质或纸质，长卵形或椭圆形，长2～6cm，先端急尖，基部钝，全缘或具疏尖锯齿，下面被硬毛，侧脉4～5对，在边缘网结；叶柄长3～5mm，密生硬毛。雌雄异株；雌花序簇生，具1～3花；雄花白色或粉红色，花瓣4～6；雌花花瓣6～8数，较雄花稍大。果球形，径4mm，熟时红色，分核常6。

产城步、通道、道县、江华、江永、宁远、南山、永兴、绥宁、宜章、资兴。生海拔200～800m山地阔叶林下或灌丛。分布于安徽南部、浙江、江西、福建、台湾、广东、海南、香港、广西、贵州。

Evergreen shrubs. Branchlets slender, slightly 4-angled, densely hirsute. Leaves membranous or papery, long ovate or elliptic, 2-6cm long, apex acute, base obtuse, margin sparsely and sharply serrulate or subentire, strigose abaxially, lateral veins 4-5 pairs, anastomosing near margin; petioles 3-5mm long, densely hirsute. Dioecious; female inflorescences fasciculate，with 1-3 flowers; male flowers pink or white，petals 4-6; female petals 6-8, slightly larger than male flowers. Fruit globose, 4mm in diam., red when mature, pyrenes 6.

Distributed in Hunan(Chengbu, Tongdao, Daoxian, Jianghua, Jiangyong, Ningyuan, Nanshan, Yongxing, Suining, Yizhang, Zixing), South Anhui, Zhejiang, Jiangxi, Fujian, Taiwan, Guangdong, Hainan, Hongkong, Guangxi and Guizhou. Grows in broad-leaved forests or thickets in mountain at alt. 200-800m.

大陆沟瓣	*Glyptopetalum continentale* (Chun et How) C. Y. Cheng et Q. S. Ma

　　常绿灌木。小枝光绿色，圆柱状或略具4棱。叶革质，椭圆形或阔椭圆形，长6～12cm，先端短渐尖或急尖，基部楔形至宽楔形，边缘有浅锯齿，略反卷，侧脉7～9对，网脉明显且在下面凸起；叶柄粗短，长4～6mm，有沟。聚伞果序簇生短枝上，短枝，多1次分枝，果序梗长1.8～2.2cm，光绿；蒴果扁球形，淡黄色或灰白色，径1～1.5cm；果皮厚而软，密被小瘤突。

　　产江永。生石灰岩山地林下。分布于广东、广西。湖南南岭地区稀见种。

　　Evergreen shrubs. Branchlets light green, terete or slightly 4-angled. Leaves leathery, elliptic or broadly elliptic, 6-12cm long, apex shortly acuminate to acute, base cuneate to broadly cuneate, shallowly serrate, slightly recurved, lateral veins 7-9 pairs, reticulate veins conspicuous and slightly raised below; petioles short, 4-6mm long, with sulcate. Cymose infructescences fascicled on short branches, branches short, usually once branched, fruit peduncle 1.8-2.2cm long, light green; capsule oblate, yellowish or gray white, 1-1.5cm in diam.; pericarp thick and soft, densely covered with tubercles.

　　Distributed in Hunan(Jiangyong), Guangdong and Guangxi. Grows in forests in limestone mountain. It is a rare species in Nanling region of Hunan.

无柄五层龙　*Salacia sessiliflora* H.-M.

常绿灌木。小枝暗灰色，具瘤状皮孔。叶薄革质，长圆状椭圆形或长圆状披针形，长10～15cm，先端渐尖或钝，基部圆或宽楔形，边缘具疏细齿，侧脉8～9对，在下面显著突起，网脉横出；叶柄长5～10mm。花少数，淡绿色，簇生叶腋，花柄极短，长不达1mm；萼片卵形，顶端钝尖，具缘毛；花瓣长圆形，顶端尖；花盘杯状；雄蕊3；子房藏于花盘内；花柱粗壮，圆锥形。浆果橙黄色至橙红色，径2～4cm。花期6月，果期10月。

产城步、江华、江永。生海拔500m以下石灰岩山地林下、沟谷边。分布于广东、广西、贵州、云南。果微甜可食。

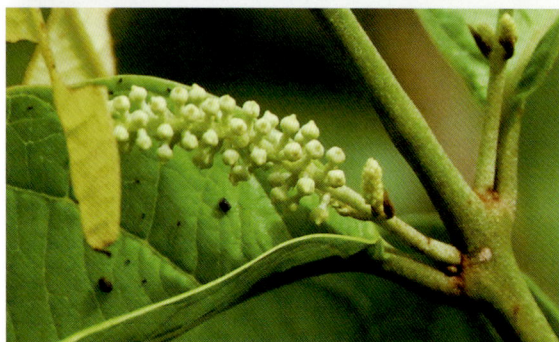

Evergreen shrubs. Branchlets dark gray, with warty lenticels. Leaves thinly leathery, oblong-elliptic or oblong-lanceolate, 10-15cm long, apex acuminate or obtuse, base rounded or broadly cuneate, margin sparsely serrulate, lateral veins 8-9 pairs, prominent rasied below, reticulate veins horizontally spreading; petioles 5-10mm long. Flowers few, light green, fascicles axillary, pedicels very short, less than 1 mm long; sepals ovate, apex obtusely pointed, ciliate; petals oblong, apex subacute; disk cupular; stamens 3; ovary concealed in disk; style stout, conical. Berry orange or red, 2-4cm in diam.. Fl. Jun, fr. Oct.

Distributed in Hunan(Chengbu, Jianghua, Jiangyong), Guangdong, Guangxi, Guizhou and Yunnan.Grows in forests, ravine sides in limestone mountain below alt. 500m. Fruit is slightly sweet and edible.

木质藤本。具粗壮卷须；幼枝褐黄色，密生糙伏毛。叶对生或近对生，长椭圆形至长圆形，长6～13cm，先端渐尖，基部圆形或楔形，侧脉4～5对，连中脉在下面显著凸起，下面沿脉散生糙伏毛；叶柄长5～10mm，被糙伏毛。雄花序腋生，长1～2cm，被糙伏毛；花黄色；萼齿5；花冠钟状，裂片5，卵形。核果长椭圆形，长2～3cm，被疏糙伏毛，熟时橙红色。花果期4～11月。

产通道、江华、江永。生海拔300～700m山地沟谷林中，常攀援树上。分布于广西、贵州、云南、广东、海南、浙江南部。越南亦产。果味甜可食。

Woody vines. With stout tendrils; young branchlets brown, densely strigose. Leaves opposite or subopposite, long elliptic to oblong, 6-13cm long, apex acuminate, base rounded or cuneate, lateral veins 4-5 pairs, lateral veins and midvein prominent abaxially, scattered strigose on veins below; petioles 5-10mm long, strigose. Male inflorescences axillary, 1-2cm long, strigose; flowers yellow; calyx teeth 5; corolla campanulate, lobes 5, ovate. Drupe long elliptic, 2-3cm long, sparsely strigose, orange-red when mature. Fl. and fr. Apr-Nov.

Distributed in Hunan(Tongdao, Jianghua, Jiangyong), Guangxi, Guizhou, Yunnan, Guangdong, Hainan and South Zhejiang, also in Vietnam. Grows in ravine forests in mountain land at alt. 300-700m. Usually climbing trees. Fruit is sweet and edible.

华南青皮木（管花青皮木） *Schoepfia chinensis* Gardn. et Champ.

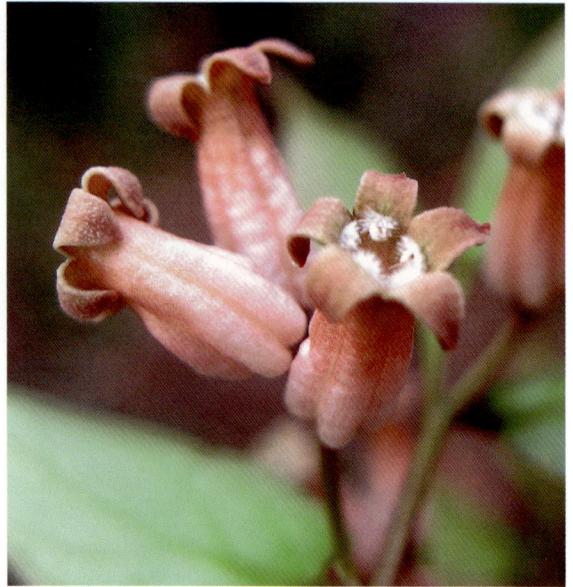

落叶小乔木。树皮紫褐色，具白色皮孔。叶厚纸质，狭椭圆形或长圆状披针形，长5～9cm，宽1.5～4cm，先端渐尖，叶脉红色，侧脉3～5对；叶柄长3～6mm。聚伞花序腋生，长2～4cm，常具3花；花无柄；花萼杯状，宿存；花冠筒状，黄白色或淡红色，柱头常短于花冠管。核果椭圆形，长1～2cm，熟时红色或紫红色。花期2～4月，果期4～6月。

产城步、新宁、道县、江华、永兴、宜章。生海拔600m以下山地阔叶林中、溪边。分布于福建、广东、广西、贵州、台湾、江西、四川、云南。

Deciduous small trees. Bark purplish brown, with white lenticels. Leaves thickly papery, narrowly elliptic or oblong-lanceolate, 5-9cm × 1.5-4cm, apex acuminate, veins red, lateral veins 3-5 pairs; petioles 3-6mm long. Cymes axillary, 2-4cm long, usually with 3 flowers; flowers sessile; calyx obcampanulate, persistent; corolla urceolate, yellowish or reddish, stigma usually shorter than corolla tube. Drupe elliptic, 1-2cm long, red or purple red when mature. Fl. Feb-Apr, fr. Apr-Jun.

Distributed in Hunan(Chengbu, Xinning, Daoxian, Jianghua, Yongxing, Yizhang), Fujian, Guangdong, Guangxi, Guizhou, Taiwan, Jiangxi, Sichuan and Yunnan. Grows in broad-leaved forests and streamsides in mountain land below alt. 600m.

红花寄生　*Scurrula parasitica* L.

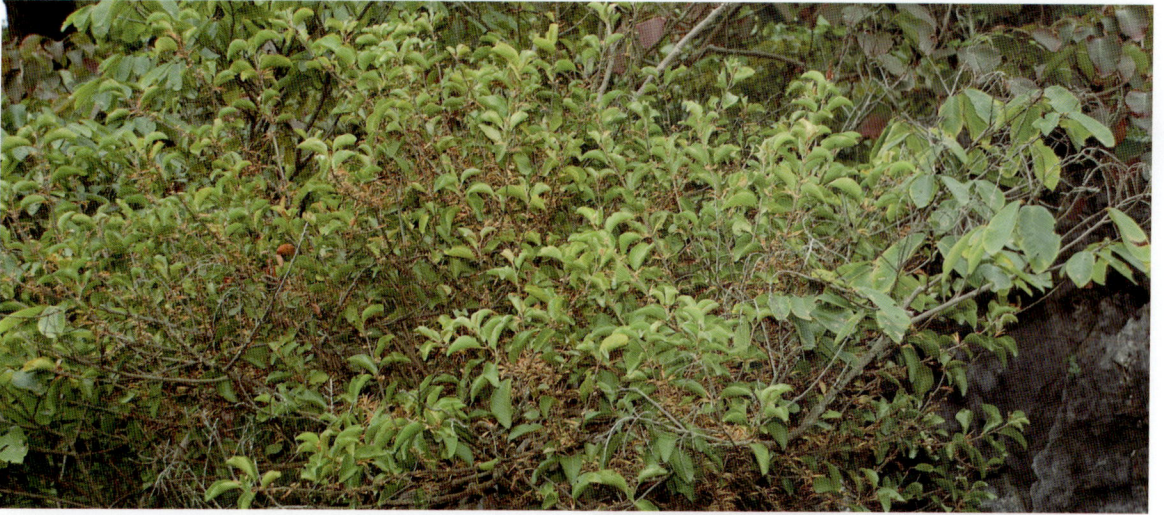

寄生灌木。小枝及嫩叶密被锈色星状毛，后脱落至无毛。叶厚纸质，卵形至长卵形，长5～8cm，先端钝，基部宽楔形，两面无毛，侧脉5～6对；叶柄长5～6mm。总状花序腋生，具3～5花，被褐色毛；总花梗长2～3mm；花梗长约2mm；花红色，花萼陀螺状，长2～2.5mm；冠檐环状；花冠蕾时管状，弯曲，下部膨胀，花冠裂片披针形，外折。果梨形，长约10mm，基部渐狭成柄，熟时红黄色。花果期9月至翌年1月。

产江永。生村边林中，寄生于无患子。分布于福建、广东、广西、贵州、海南、江西、四川、台湾、西藏、云南。东南亚亦产。

Parasitic shrubs. Young twigs and leaves densely ferruginous stellate-pubescent, later gradual glabrous. Leaves thickly papery, ovate to long-ovate, 5-8cm long, apex obtuse, base broadly cuneate, glabrous on both surfaces, lateral veins 5-6 pairs; petioles 5-6mm long. Racemes axillary, with 3-5 flowers, brown hairs; peduncle 2-3mm long; pedicels ca.2mm; flowers red, Calyx turbinate, 2-2.5mm long, limb annular; corolla bud tubular, curved, basal part inflated, corolla lobes lanceolate, reflexed. Fruit pyriform, ca.10mm long, base tapering into stalk, reddish yellow when mature. Fl. and fr. Sep to Jan of 2nd year.

Distributed in Hunan(Jiangyong), Fujian, Guangdong, Guangxi, Guizhou, Hainan, Jiangxi, Sichuan, Taiwan, Tibet and Yunnan, also in Southeast Asia. Grows in forests of village edge, parasitic in *Sapindus mukorossi*.

广寄生（桑寄生） *Taxillus chinensis* (DC.) Danser

寄生灌木。嫩枝、芽、幼叶均密被黄褐色星状毛，后渐无毛。叶厚纸质，卵形至长卵形，长3～6cm，先端圆钝，基部楔形或阔楔形，侧脉3～4对，略明显；叶柄长8～10mm。伞形花序单生或孪生，有时生于落叶小枝，具1～4花，常2花，花序和花被星状毛；花褐色，花托椭圆状或卵球形；花冠花芽时管状，长2.5～2.7cm，稍弯，下部膨胀，顶部卵球形，裂片4枚，反折。果浅黄色，椭圆形或近球形，长8～10mm，密被瘤突，熟时变平滑。花果期4～12月。

产通道、新宁、江华。生海拔500m以下阔叶林中，寄生油茶、油桐树上。分布于福建、广东、广西、海南、香港、澳门。东南亚亦产。

Parasitic shrubs. Young twigs, buds, young leaves densely tawny stellate hairs, later becoming glabrous. Leaves thickly papery, ovate to long ovate, 3-6cm long, apex obtuse, base cuneate or broadly cuneate, lateral veins 3-4 pairs, slightly conspicuous; petioles 8-10mm long. Umbels solitary or 2 together, sometimes at leafless nodes, with 1-4 flowers, usually 2 flowers, inflorescences and flowers with stellate-pubescent; flowers brown, torus elliptic or ovoid; corolla tubular when flower buds, 2.5-2.7cm long, slightly curved, basal part inflated, ovoid, lobes 4, reflexed. Fruit yellowish, elliptic or subglobose, 8-10mm long, densely verrucose, becoming smooth when mature. Fl. and fr. Apr-Dec.

Distributed in Hunan(Tongdao, Xinning, Jianghua), Fujian, Guangdong, Guangxi, Hainan, Hongkong and Macao, also in Southeast Asia. Grows in broad-leaved forests below alt. 500m, parasitic in *Camellia oleifera*, *Vernicia fordii*.

木兰寄生　*Taxillus limprichtii* (Gruning) H. S. Kiu

寄生灌木。嫩芽密被黄褐色星状毛，后变无毛。叶革质，对生或近对生，卵状长圆形或倒卵形，长4～12cm，先端钝或近圆形，基部楔形，两面无毛，侧脉4～5对；叶柄长5～12mm。伞形花序腋生，具3～5花，被星状毛；花红色或橙色，花托长卵状，长1.5～2.5mm，副萼环状；花冠花芽时管状，长2.7～3cm，顶部长圆形，开花时顶部4裂，外折。果长圆形，被疏毛，具小瘤体，熟时浅黄色，长约7mm，果皮不平坦，无毛。花期10月至翌年3月，果期6～7月。

产通道、城步、宜章。生低山阔叶林中，寄主为枫香。分布于福建、广东、广西、贵州、江西南部、四川、台湾、云南。

Parasitic shrubs. Young buds densely yellow-brown stellate hairs, becoming glabrous. Leaves leathery, opposite or subopposite, ovate-oblong or obovate, 4-12cm long, apex obtuse or suborbicular, base cuneate, glabrous on both surfaces, lateral veins 4-5 pairs; petioles 5-12mm long. Umbels axillary, with 3-5 flowers, stellate-pubescent; flowers red or orange, torus long-ovate, 1.5-2.5mm long, epicalyx annular; corolla tubular when flower buds, 2.7-3cm long, top oblong, apically 4-lobed when blossom, reflexed. Fruit oblong, sparsely hairy, with tubercles, light yellow when mature, ca.7mm long, pericarp not smooth, glabrous. Fl. Oct to Mar of 2nd year, fr. Jun-Jul.

Distributed in Hunan(Tongdao, Chengbu, Yizhang), Fujian, Guangdong, Guangxi, Guizhou, South Jiangxi, Sichuan, Taiwan and Yunnan. Grows in broad-leaved forests in low mountain, parasitic in *Liquidambar formosana*.

长约1mm；苞片淡红色，长12～22mm，顶端渐尖，基部圆钝或浅心形；花红色或橙色；花冠管上半部膨胀，具5棱，裂片狭长圆形，反折。果椭圆形，长8～12mm，熟时黄色，被星状毛。花期4～7月，果期8～10月。

产新宁、城步、绥宁、江永、零陵、宜章。生沟谷阔叶林中，寄生于枥类或石楠属植物上。分布于福建、广东、广西、贵州、江西。

寄生灌木。幼枝密被黄褐色或锈色星状毛，后脱落至无毛。叶薄革质，互生或近对生，或3～4枚簇生短枝上，长圆形或长卵形，长3～7cm，先端急尖或钝，基部楔形或圆钝；叶柄长3～7mm，无毛。花序1～3个腋生，具3～5花，总花梗长7～12mm，花梗

Parasitic shrubs. Young twigs densely yellow-brown or rust-colored stellate hairs, later gradual glabrous. Leaves thinly leathery, alternate or subopposite, or 3-4 fascicled on short branches, oblong or long ovate, 3-7cm long, apex acute or obtuse, base cuneate or obtuse; petioles 3-7mm long, glabrous. Inflorescences 1-3 axillary, with 3-5 flowers, peduncle 7-12mm long, pedicels ca.1mm long; bracts light red, 12-22mm long, apex acuminate, base rounded or shallowly cordate; flowers red or orange; corolla tube apical portion inflated, 5-angled, lobes narrowly oblong, reflexed. Fruit elliptic, 8-12mm long, yellow when mature, stellate hairs. Fl. Apr-Jul, fr. Aug-Oct.

Distributed in Hunan(Xinning, Chengbu, Suining, Jiangyong, Lingling, Yizhang), Fujian, Guangdong, Guangxi, Guizhou and Jiangxi. Grows in ravine broad-leaved forests, parasitic in *Quercus* or *Photinia* plants.

翼核果 *Ventilago leiocarpa* Benth.

攀援灌木。茎灰褐色，具条纹，无毛。叶薄革质，卵形或卵状椭圆形，长4~8cm，先端渐尖，基部圆形，全缘或具浅圆齿，两面无毛，侧脉4~6对，网脉明显；叶柄长3~5mm。腋生聚伞花序或有时成顶生圆锥花序；花小，绿白色，5数；萼片三角形；花瓣匙形；雄蕊5；子房藏花盘内。核果球形，顶端具矩圆形的翅，翅宽7~9mm，基部具宿存的萼筒。

产江永、汝城、通道。生山地沟谷疏林下或灌丛中。分布于福建、广东、广西、香港、台湾、云南、贵州。越南、缅甸、印度、非洲热带地区亦产。

Climbing shrubs. Stem gray brown, with stripes, glabrous. Leaves thinly leathery, ovate or oblong-ovate, 4-8cm long, apex acuminate, base rounded, margin entire or shallowly crenate, glabrous on both surfaces, lateral veins 4-6 pairs, reticulate veins conspicuous; petioles 3-5mm long. Cymes axillary or sometimes arranged in terminal panicles; flowers small, green-white, 5-merous; calyx triangular; petals spatulate; stamens 5; ovary immersed in disk. Drupe globose, apex with oblong wings, wing 7-9mm wide, with persistent calyx tube at base.

Distributed in Hunan(Jiangyong, Rucheng, Tongdao), Fujian, Guangdong, Guangxi, Hongkong, Taiwan, Yunnan and Guizhou, also in Vietnam, Burma, India, Africa tropical regions. Grows in sparse forests or shrubs in mountain ravine.

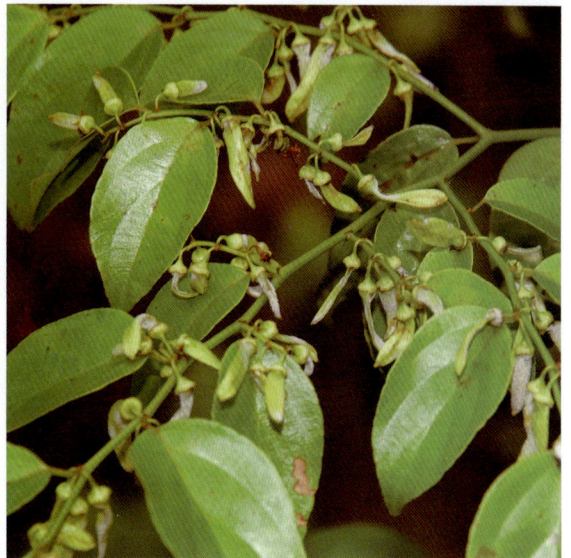

常绿灌木。叶具5～15小叶，小叶片卵形到披针形，长4～10cm，宽2～5cm，先端锐尖或渐尖，基部不对称，边缘具锯齿，稀波状，两面无毛，下面被凸起的油点；小叶柄长4～8mm。花序顶生；花4数；萼裂片细小；花瓣长圆形，长3～4mm；雄蕊8；子房圆球形；花柱短于子房。果球形，熟时蓝黑色，径1～1.5cm；种子1～2。花期6～7月，果期9～11月。

产新宁、东安、江永、江华、临武、宜章、宁远。生低海拔石灰岩山地。分布于广东、广西、贵州、四川、云南。越南亦产。

Evergreen shrubs. Leaves 5-15-foliolate, leaflet blades ovate to lanceolate, 4-10cm × 2-5cm, apex acute to acuminate, base asymmetric, margin serrate or rarely repand, glabrous on both surfaces; leaflet petiolules 4-8mm. Inflorescences terminal; flowers 4-merous, calyx lobes small; petals oblong, 3-4mm long; stamens 8; ovary globose; style shorter than ovary. Fruit globose, bluish black when mature, 1-1.5cm in diam., seeds 1-2. Fl. Jun-Jul, fr. Sep-Nov.

Distributed in Hunan(Xinning, Dongan, Jiangyong, Jianghua, Linwu, Yizhang, Ningyuan), Guangdong, Guangxi, Guizhou, Sichuan and Yunnan, also in Vietnam. Grows in limestone mountain at low altitude.

常绿灌木或小乔木。老枝灰白色或淡黄灰色。叶具2～5小叶，小叶片卵形或卵状披针形，长2～9cm，宽1.5～6cm，先端渐尖至短尖，基部楔形，全缘或具细圆齿，侧脉4～8对；小叶柄短于1cm。花序顶生及腋生；花5数；萼片卵形到披针形，长达2mm，在果期宿存；花瓣白色，狭椭圆形到倒披针形，长2cm；雄蕊10。果橙黄或朱红色，狭椭圆形，稀卵形，长1～2cm。种子具长柔毛。花期4～10月，果期9～12。

产通道、江华、江永、道县、宜章。生低海拔石灰岩山地林下或灌丛。分布于福建、广东、广西、贵州南部、海南、台湾、云南。东南亚、澳大利亚亦产。

Evergreen shrubs or small trees. Old branches grayish white or pale yellowish gray. Leaves 2-5-foliolate; leaflet blades ovate or ovate-lanceolate, 2-9 × 1.5-6cm, margin entire or crenulate, apex acuminate to mucronate; leaflet petiolules less than 1 cm. Inflorescences terminal and axillary; flowers 5-merous; sepals ovate to lanceolate, up to 2mm long, persistent in fruit; petals white, narrowly elliptic to oblanceolate, 2cm long; stamens 10. Fruit orange to vermilion, narrowly ellipsoid or rarely ovoid, 1-2cm long; seeds villous. Fl. Apr-Oct, fr. Sep-Dec. Distributed in Hunan(Tongdao, Jianghua, Jiangyong, Daoxian, Yizhang), Fujian, Guangdong, Guangxi, South Guizhou, Hainan, Taiwan and Yunnan, also in Southeast Asia, Australia. Grows in forests in limestone mountain or thickets.

麻 楝 *Chukrasia tabularis* A. Juss.

落叶乔木。小枝红褐色，无毛，具白色皮孔。叶常为偶数羽状复叶，具10～16小叶；叶柄圆柱形，长4～7cm；小叶纸质，互生，卵形至矩圆状披针形，长7～12cm，无毛。聚伞圆锥花序顶生，松散，长约为叶的一半，疏散；花黄色带紫；萼杯状，裂片4～5；花瓣4～5，矩圆形；雄蕊花丝合生成筒。蒴果近球形，径3.5～4cm，常3瓣裂，木质，表面粗糙和疣状。花期4～5月，果期7月至翌年1月。

产道县、江永。生山坡、山谷疏林中。分布于广东、广西、贵州、云南、西藏、浙江南部。越南至印度亦产。木材坚硬芳香，为家具、胶合板、雕刻的好材料。

Deciduous trees. Branchlets reddish-brown, glabrous, with white lenticels. Leaves usually paripinnate, with 10-16 leaflets; petioles cylindric, 4-7cm long; leaflets papery, alternate, ovate to oblong-lanceolate, 7-12cm long, glabrous. Thyrses terminal, lax, ca.1/2 as long as leaves; flowers yellow with purple; calyx cupular, lobes 4-5; petals 4-5, oblong; stamens filaments connate into tube. Capsule subglobose, 3.5-4cm in diam., usually 3-valved, woody, surface coarse and verrucose. Fl. Apr-May, fr. Jul to Jan of 2nd year.

Distributed in Hunan(Daoxian, Jiangyong), Guangdong, Guangxi, Guizhou, Yunnan, Tibet and South Zhejiang, also in Vietnam to India. Grows on mountain slopes, and in sparse forests in hill valley. The fragrant hard timber is valuable for making furniture, plywood and carving planks.

黄梨木 Boniodendron minus (Hemsl.) T. Chen

落叶小乔木。小枝被短柔毛。偶数羽状复叶，长9～12cm，具10～20小叶；叶柄纤细，长1～2cm，小叶纸质，披针形或椭圆形，长2～3cm，先端钝，基部偏斜，边缘具钝锯齿；小叶柄极短，长约1mm。聚伞圆锥花序顶生；花淡黄色至近白色；萼片5，外面被白色短柔毛；花瓣长圆形，具羽状脉纹，外面被白色疏柔毛；雄蕊8枚；子房具3沟槽，被毛。蒴果近球形，具3翅，径约2cm；种子黑褐色。花期5～6月，果期7～8月。

产道县、嘉禾、宁远、江华、江永、宜章。生海拔200～600m石灰岩山地疏林或密林中。分布于广东、广西、贵州、云南。用材树种。

Deciduous small trees. Branchlets pubescent. Paripinnate, 9-12cm long, with 10-20 leaflets; petioles slender, 1-2cm long, leaflets papery, lanceolate or elliptic, 2-3cm long, apex obtuse, base oblique, margin obtusely serrate; petiolules very short, ca.1mm long. Thyrses terminal; flowers pale yellow to nearly white; sepals 5, outside white pubescent; petals oblong, pinnately veined, sparsely white pilose; stamens 8; ovary 3-furrowed, hairy. Capsule subglobose, with 3 wings, ca.2cm in diam.; seeds black brown. Fl. May-Jun, fr. Jul-Aug.

Distributed in Hunan(Daoxian, Jiahe, Ningyuan, Jianghua, Jiangyong, Yizhang), Guangdong, Guangxi, Guizhou and Yunnan. Grows in sparse forests or dense forests in limestone mountain at alt. 200-600m. This species is used for timber.

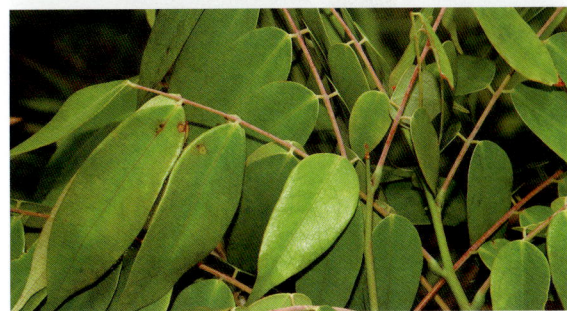

于叶腋，长2.5～5cm，总花梗和花梗纤细；花白色、淡黄色或淡红色，径5～7mm；萼片5，花瓣5。蓇葖果椭圆形或斜卵形，熟时红色。花期3～9月，果期5月至翌年3月。

产汝城。生低海拔沟谷疏林下。分布于福建、广东、广西、云南。越南、斯里兰卡、印度、印度尼西亚亦产。湖南新记录科、属、种。

Lianoid Shrubs. Odd pinnae, usually with 7-17 leaflets, leaflets subleathery, ovate to ovate oblong, 2-4cm long, apex acuminate and obtuse, base oblique, margin entire, glossy adaxially and slightly glaucous-green abaxially; petiolules short, 2mm long, glabrous. Panicles, clustered in leaf axils, 2.5-5cm long, peduncle and pedicels slender; flowers white, pale yellow or pale red, 5-7mm in diam.; sepals 5, petals 5. Follicles elliptic or obliquely ovate, red when mature. Fl. Mar-Sep, fr. May to Mar of 2nd year.

Distributed in Hunan(Rucheng), Fujian, Guangdong, Guangxi and Yunnan, also in Vietnam, Sri Lanka, India, Indonesia. Grows in ravine sparse forests at low altitude. New recorded family, genus, species in Hunan province.

藤状灌木。奇数羽状复叶，常具7～17小叶，小叶近革质，卵形至卵状矩圆形，长2～4cm，先端渐尖而钝，基部偏斜，全缘，无毛，上面光亮，下面稍带粉绿色；小叶柄极短，长2mm，无毛。圆锥花序，丛生

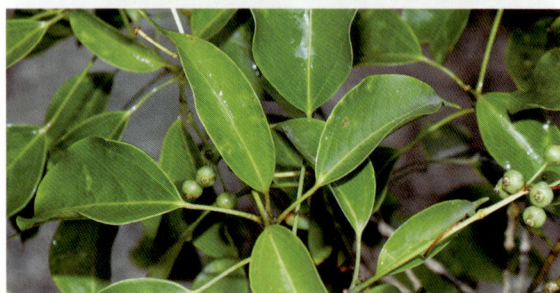

显；叶柄纤细，长达9cm。伞形花序4～5个组成顶生的复伞形花序；总花梗长1.5～2cm；花梗长约5mm；花瓣5；雄蕊5；子房下位。果球形，径约8mm，略具棱，熟时暗紫色。花期6～7月，果期10月。

　　产道县、江永、宁远、宜章。生山谷密林或山坡疏林中。分布于广东、广西、贵州、海南、云南。

Evergreen trees. Leaves leathery, glabrous, not glandular punctate, elliptic, oblong or ovate-elliptic, 6-11cm long, apex long acuminate to caudate, base cuneate, margin entire, pinnately veined, lateral veins 5-8 pairs, reticulate veins inconspicuous or conspicuous; petioles slender, up to 9cm long. Component umbels, terminal, with 4-5 umbels; peduncle 1.5-2cm long; pedicels ca.5mm long; petals 5; stamens 5; ovary inferior. Fruit globose, ca.8mm in diam., slightly ribbed, dark purple when mature. Fl. Jun-Jul, fr. Oct.

Distributed in Hunan(Daoxian Jiangyong, Ningyuan, Yizhang), Guangdong, Guangxi, Guizhou, Hainan and Yunnan. Grows in dense forests in valley or in sparse forests on mountain slopes.

　　常绿乔木。叶革质，无毛，无腺体，椭圆形、长圆形或卵状椭圆形，长6～11cm，先端长渐尖或尾尖，基部楔形，全缘，羽状脉，侧脉5～8对，网脉不明显或明

常绿灌木。叶革质、纸质或薄纸质，无毛，常无腺体，2型，不裂或掌状深裂；不裂叶椭圆形至线状披针形，长2.5~12cm，基出3脉，有时不甚明显，侧脉常5~9对，两面网脉均不明显；分裂叶倒三角形，掌状3深裂。伞形花序单生或2~3个聚生；花绿色；花瓣4~5；雄蕊4~5；子房下位。果球形，平滑无棱，径5~6mm。花期8~9月，果期9~10月。

产通道、江华、江永、蓝山、炎陵、新宁、宜章、资兴。生山地灌丛或沟谷阔叶林下。分布于福建西南部、广东、广西、香港、江西、云南。

Evergreen shrubs. Leaves leathery, papery or thinly papery, glabrous, usually not glandular punctate, dimorphic, undivided or palmatipartite; unlobed blades elliptic to linear-lanceolate, 2.5-12cm long, basal veins 3, sometimes not very obvious, lateral veins usually 5-9 pairs, reticulate veins inconspicuous on both surfaces; lobed blades obtriangular, palmately deeply 3-cleft. Umbels solitary or 2-3-fascicled; flowers green; petals 4-5; stamens 4-5; ovary inferior. Fruit globose, smooth, without ribs, 5-6mm in diam.. Fl. Aug-Sep, fr. Sep-Oct.

Distributed in Hunan(Tongdao, Jianghua, Jiangyong, Lanshan, Yanling, Xinning, Yizhang, Zixing), Southwest Fujian, Guangdong, Guangxi, Hongkong, Jiangxi and Yunnan. Grows in mountain shrubs or in ravine broad-leaved forests.

常绿小乔木。嫩枝、叶轴、叶柄、花序轴、花梗、花萼均密被暗锈色茸毛。叶聚生茎顶，形似伞，4～5回羽状复叶，长达90cm，小叶纸质，椭圆形至狭椭圆形，长2～8cm，先端渐尖，基部宽楔形，两面无毛，小叶无柄或具短柄；叶柄粗壮，长15～40cm。由伞形花序再组成圆锥花序顶生，长达70cm；伞形花序径1～1.5cm，具多花；总花梗长1～2cm；花梗长1.5～2.5mm；花淡黄白色；花瓣5；雄蕊5。果实扁球形，径7～8mm，熟时黑色，宿存花柱长2～3mm。花期11～12月，果期翌年1～2月。

产通道、江华、汝城。生海拔500m以下山地沟谷湿润林中。分布于福建、广东、广西、江西。

Evergreen small trees. Young branchlets, leaf rachises, inflorescence rachis, petioles, pedicels, calyx densely dark ferruginous tomentose. Leaves fascicled at stem apices, like umbelliferae, 4-5 pinnae, up to 90cm long, leaflets papery, elliptic to narrowly elliptic, 2-8cm long, apex acuminate, base broadly cuneate, glabrous on both surfaces, leaflets sessile or short petiolate；petioles stout, 15-40cm long. Umbels arranged in panicles, terminal, up to 70cm long; umbels 1-1.5cm in diam., with many flowers; peduncle 1-2cm long; pedicels 1.5-2.5mm long; flowers yellowish white; petals 5; stamens 5. Fruit oblate, 7-8mm in diam., black when mature, persistent style 2-3mm long. Fl. Nov-Dec, fr. Jan-Feb of 2nd year.

Distributed in Hunan(Tongdao, Jianghua, Rucheng), Fujian, Guangdong, Guangxi and Jiangxi. Grows in moist forests in mountain ravine below alt. 500m.

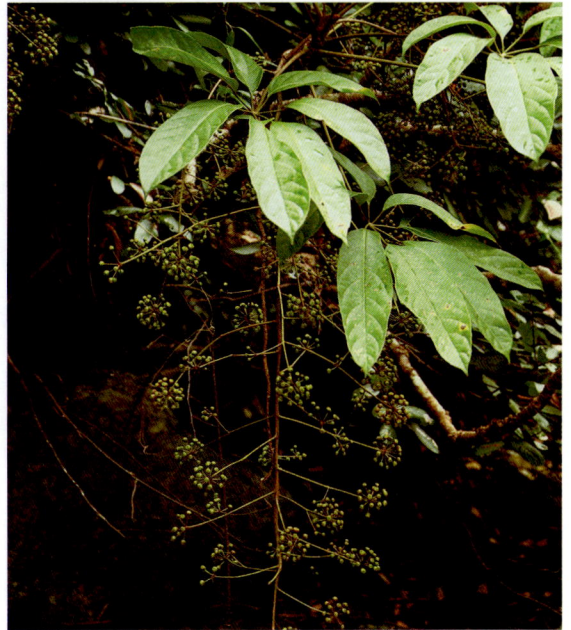

常绿小乔木。小枝密被黄棕色星状茸毛，不久即脱净。掌状复叶，具7～15小叶；小叶纸质至薄革质，长椭圆状披针形至卵状披针形，长10～25cm，上面无毛，下面密生灰色星状茸毛；小叶柄极不等长，中央的长3～13cm，两侧的长1～2cm。由伞形花序排成圆锥花序顶生，花序密被黄棕色星状短毛；总花梗长2～3.5cm；花瓣5；雄蕊5；子房下位，5室，花柱合生成柱状。果球形，具5棱，径约4mm，宿存花柱长约2mm。花期9月，果期10～11月。

产通道、道县、江永、宜章。生山地沟谷湿林中。分布于福建、广东、广西、贵州、江西、云南、浙江南部。

Evergreen small trees. Branchlets densely yellowish-brown stellate tomentose, soon gradual glabrous. Leaves palmately, with 7-15 leaflets; leaflets papery to thinly leathery, long elliptic-lanceolate to ovate-lanceolate, 10-25cm long, glabrous adaxially, densely gray stellate tomentose abaxially; petiolules very unequal in length, central 3-13cm long, both sides 1-2cm long. Umbels arranged in panicles, terminal, inflorescence densely yellowish-brown stellate pubescent; peduncle 2-3.5cm long; petals 5; stamens 5; ovary inferior, 5-loculed, styles united into a column. Fruit globose, with 5-ribbed, ca.4mm in diam., persistent style ca.2mm long. Fl. Sep, fr. Oct-Nov.

Distributed in Hunan(Tongdao, Daoxian, Jiangyong, Yizhang), Fujian, Guangdong, Guangxi, Guizhou, Jiangxi, Yunnan and South Zhejiang. Grows in moist forests in mountain ravine.

铁榄 (山胶木)   *Sinosideroxylon pedunculatum* (Hemsl.) H. Chuang

常绿小乔木。小枝、幼叶密被锈色柔毛，老枝密被皮孔。叶革质，互生，常密聚小枝顶端，卵形或卵状披针形，长6～9cm，先端渐尖，基部楔形，两面无毛，侧脉8～12对，近平展，连网脉在两面均明显；叶柄长1～3cm。总状花序长2～2.5cm，花序梗长1～3cm，被锈色微柔毛，具多花；花冠浅黄色。浆果卵球形，长2.5cm，熟时黑色。

产通道、道县、江永、江华。生石灰岩低山阔叶林中。分布于广东、广西、云南。越南亦产。

Evergreen small trees. Branchlets, young leaves densely rust colored pubescent, old branches densely lenticels. Leaves leathery, alternate, usually clustered at the top of branchlets, ovate or ovate-lanceolate, 6-9cm long, apex acuminate, base cuneate, glabrous on both surfaces, lateral veins 8-12 pairs, horizontally spreading, and with reticulate veins conspicuous on both surfaces; petioles 1-3cm long. Racemes 2-2.5cm long, peduncle 1-3cm long, rusty puberulous, with many flowers; corolla pale yellow. Berry ovoid, 2.5cm long, black when mature.

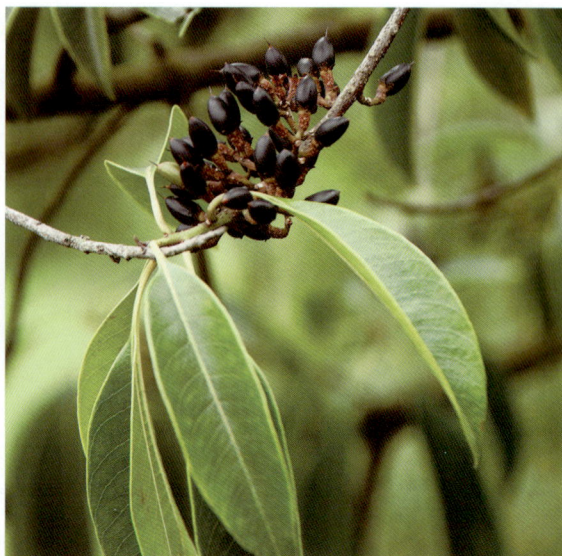

Distributed in Hunan(Tongdao, Daoxian, Jiangyong, Jianghua), Guangdong, Guangxi and Yunnan, also in Vietnam. Grows in broad-leaved forests in low limestone mountain.

常绿小灌木。茎纤细，具匍匐茎。叶坚纸质，卵状椭圆形，长1.5～3.5cm，先端钝尖或急尖，基部楔形，边缘具浅波状齿和腺体，无毛，下面被腺状柔毛，侧脉4～5对，不连成边脉；叶柄长2～5mm。伞形花序，侧生，花枝长2～4cm，具小形叶；花瓣粉红色。果球形，径约7mm，熟时红色，无腺体。花期5～7月，果期10～12月。

产新宁、城步、江华、宜章。生海拔500m以下石灰岩林下。分布于广东、广西、贵州、四川。

Evergreen small shrubs. Stem slender, with creeping rhizomes. Leaves hard papery, ovate-elliptic, 1.5-3.5cm long, apex obtuse or acute, base cuneate, margin shallowly undulate-dentate and glands, glabrous, glandular pilose below, lateral veins 4-5 pairs, without intramarginal veins; petioles 2-5mm long. Umbels, lateral, blooming branch 2-4cm long, with small shaped leaf; petals pink. Fruit globose, ca.7mm in diam., red when mature, without glands. Fl. May-Jul, fr. Oct-Dec.

Distributed in Hunan(Xinning, Chengbu, Jianghua, Yizhang), Guangdong, Guangxi, Guizhou and Sichuan. Grows in limestone forests below alt. 500m.

亚伞形花序或伞房花序，侧生，密被微柔毛；总梗长3～6cm；花梗长6～10mm，常带红色；花瓣白色，稀粉红色。果球形，径5～6mm，熟时红色，具腺点。花期6～7月，果期10～12月。

产通道、江华、蓝山、宜章。生山谷密林下。分布于广东、广西、贵州、四川。

常绿小灌木。茎纤细，具匍匐茎。叶近革质，长圆状披针形或卵状披针形，长4～6.5cm，先端渐尖，基部钝至圆形，边缘具浅圆齿及腺点，无毛，下面散布腺点，侧脉不明显；叶柄长5～8mm，具沟。

Evergreen small shrubs. Stem slender, with creeping rhizomes. Leaves subleathery, oblong-lanceolate or ovate-lanceolate, 4-6.5cm long, apex acuminate, base obtuse to rounded, margin with shallowly crenate and glandular punctate, glabrous, scattered glands abaxially, lateral veins inconspicuous; petioles 5-8mm long, with sulcate. Subumbels or corymbs, lateral, densely puberulent; peduncle 3-6cm long; pedicels 6-10mm long, usually reddish; petals white, rarely pink. Fruit globose, 5-6mm in diam., red when mature, with glands. Fl. Jun-Jul, fr. Oct-Dec.

Distributed in Hunan(Tongdao, Jianghua, Lanshan, Yizhang), Guangdong, Guangxi, Guizhou and Sichuan. Grows in dense forests in valleys.

常绿半灌木，具匍匐根状茎。茎高达45cm，幼嫩时被细微鳞片，后脱落。叶纸质，互生，倒卵形或椭圆形，长3～5cm，先端钝，基部楔形，边缘中部以上具波状齿或近全缘，下面密布细小褐色鳞片，侧脉8～10对，至叶缘连结成边脉。伞形花序腋生，总花梗和花梗被褐色鳞片；花冠裂片卵形，无腺体。果球形，径5mm，红色，后变黑色，无腺体，无毛。花期4～6月，果期10～12月。

产通道、新宁、临武、宜章。生山地沟谷林下。分布于福建、广东、广西、江西、四川、台湾、浙江南部。

Evergreen subshrubs, with creeping rhizomes. Stem up to 45 cm tall, minutely scaly when young, glabrescent. Leaves papery, alternate, obovate or elliptic, 3-5cm long, apex obtuse, base cuneate, margin undulate-dentate or nearly entire above middle, densely small brown scaly below, lateral veins 8-10 pairs, anastomosing near margin. Umbels axillary, peduncle and pedicels brown scaly; corolla lobes ovate, without glands. Fruit globose, 5mm in diam., red, later becoming black, without glands, glabrous. Fl. Apr-Jun, fr. Oct-Dec.

Distributed in Hunan(Tongdao, Linwu, Xinning, Yizhang), Fujian, Guangxi, Jiangxi, Guangdong, Sichuan, Taiwan and South Zhejiang. Grows in forests in mountain ravines.

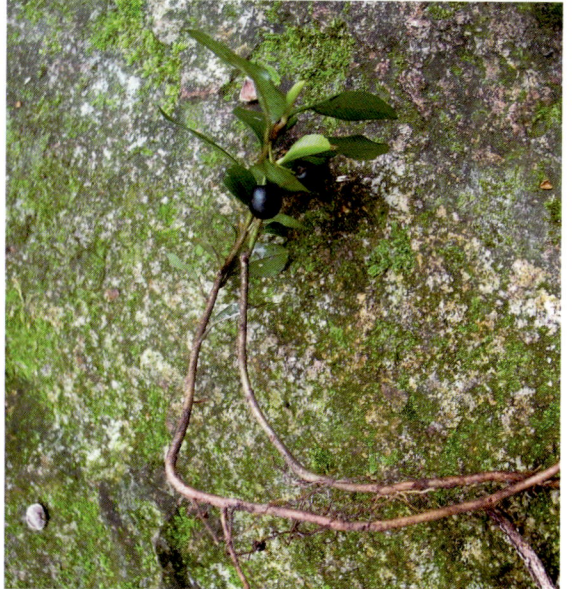

郎伞木（美丽紫金牛） *Ardisia elegans* Andr.

常绿灌木。叶坚纸质，长圆状椭圆形，长9～13cm，先端急尖或渐尖，基部楔形，边缘具凸波状圆齿和大型腺点，无毛，下面干时淡褐色，无鳞片，腺点不明显，侧脉12～15对，在边缘连结；叶柄1～1.5cm，有窄翅。复伞形花序顶生；萼片卵形，先端急尖或钝，无腺点；花瓣粉红色，稀白色，卵形，无腺点。果球形，径约1cm，红色，具泡状腺点。花期6～7月，果期10月。

产道县、江永。生海拔300～600m山地溪边。分布于广东、广西、海南。越南亦产。

Evergreen shrubs. Leaves hard papery, oblong-elliptic, 9-13cm long, apex acute or acuminate, base cuneate, margin convex undulate-crenate dentate and large glandular punctate, glabrous, pale brown when dry, not scaly, glands inconspicuous, lateral veins 12-15 pairs, anastomosing near margin; petioles 1-1.5cm long, with narrow wing. Compound umbels terminal; sepals ovate, apex acute or obtuse, without glands; petals pink, rarely white, ovate, without glands. Fruit globose, ca.1cm in diam., red, with vesicular glands. Fl. Jun-Jul, fr. Oct.

Distributed in Hunan(Tongdao, Jiangyong), Guangdong, Guangxi and Hainan, also in Vietnam. Grows in streamsides in mountain land at alt. 300-600m.

常绿半灌木，具匍匐木质根状茎。幼枝、叶柄、幼叶、花序密被红棕色卷曲毛。叶常集生枝顶，纸质，椭圆形或倒卵形，长7～14cm，先端急尖或钝，边缘浅波状，具缘毛，两面被腺体和褐色卷曲毛。伞形花序具7～15花，侧生枝顶；花白色或少粉红色；花瓣卵形，被黑腺体。果球形，径6～7mm，无毛，熟时鲜红色。花期6～7月，果期12月至翌年2月。

产江华、江永、通道、汝城。生山地、山谷林下。分布于福建、广东、广西、贵州、海南、四川、云南。株形雅致，枝叶红色，可栽培观赏。

Evergreen subshrubs, with creeping woody rhizomes. Young branchlets, petioles, young leaves, inflorescences densely red-brown crisped hairs. Leaves usually crowded at the top of branchlets, papery, elliptic or obovate, 7-14cm long, apex acute or obtuse, margin sinuolate, ciliate, with glands and brown crisped hairs on both surfaces. Umbels, lateral at the top of branchlets, with 7-15 flowers; Flowers whitish or rarely pinkish; Petals ovate, with black glands. Fruit globose, 6-7mm in diam., glabrous, cardinal red when mature. Fl. Jun-Jul, fr. Dec to Feb of 2nd year.

Distributed in Hunan(Jianghua, Jiangyong, Tongdao, Rucheng), Fujian, Guangdong, Guangxi, Guizhou, Hainan, Sichuan and Yunnan. Grows in forests in mountain land and valleys. It can be cultivated to watch for its elegant tree-shape and red leaves.

莲座紫金牛　　*Ardisia primulaefolia* Gardn. et Champ.

常绿小灌木。叶莲座状丛生，椭圆形，先端钝或突尖，基部圆形，长6～14cm，边缘具波状圆齿，具边缘腺体，两面有时紫红色，被卷曲锈色长柔毛及长缘毛，侧脉5～7对，近叶缘网结，明显；叶柄长5～10mm，密被长柔毛。聚伞花序，总梗长3～8cm；花瓣粉红色，广卵形，具腺体。果球形，径4～6mm，熟时红色，疏具腺体。花期6～7月，果期11～12月。

产江永、汝城。生山坡潮湿林下。分布于云南、广西、广东、江西、福建。越南亦产。可栽培供观赏。

Evergreen small shrubs. Leaves in a rosette, elliptic, apex obtuse or abruptly acuminate, base rounded, 6-14cm long, margin undulate crenate, with marginal glands, sometimes purplish-red on both surfaces, curly rusty villous and long ciliate, lateral veins 5-7 pairs, anastomosing near margin, conspicuous; petioles 5-10mm long, densely villous. Cymes, peduncle 3-8cm long; petals pink, broadly ovate, with glands. Fruit globose, 4-6mm in diam., red when mature, sparsely glands. Fl. Jun-Jul, fr. Nov-Dec.

Distributed in Hunan(Jiangyong, Rucheng), Yunnan, Guangxi, Guangdong, Jiangxi and Fujian, also in Vietnam. Grows in humid forest on mountain slopes. It can be planted for ornament.

　　常绿灌木。小枝纤细，被锈色鳞片。叶纸质，长圆状椭圆形或倒披针形，长7～10cm，先端渐尖，基部楔形，全缘呈微波状，无毛，下面被暗褐色鳞片，侧脉多数而密，近横出，边缘网结为边脉，无边缘腺体。花序近伞状，侧生或腋生，长1～6cm；萼裂片卵形，无腺体；花瓣宽椭圆形，白色或粉红色。果扁球形，具5钝棱，径5～7mm，无腺体。

　　产江永。生山地林中或林缘。分布于福建、广东、广西、贵州、海南、四川、台湾、云南。日本、马来西亚亦产。湖南南岭地区稀见。

Evergreen shrubs. Branchlets slender, rusty scaly. Leaves papery, oblong-elliptic or oblanceolate, 7-10cm long, apex acuminate, base cuneate, margin entire and sinuolate, glabrous, dark brown scaly abaxially, lateral veins numerous and dense, nearly horizontally spreading, anastomosing near margin, without marginal glands. Inflorescences subumbellate, lateral or axillary, 1-6cm long; calyx lobes ovate, without glands; Petals broadly elliptic, white or pink. Fruit oblate, with 5 obtuse angles, 5-7mm in diam., without glands.

　　Distributed in Hunan(Jiangyong), Fujian, Guangdong, Guangxi, Guizhou, Hainan, Sichuan, Taiwan and Yunnan, also in Japan, Malaysia. Grows in forests or forest margins in mountain land. It is a rare species in Nanling region of Hunan.

常绿攀援灌木。小枝具纵棱。叶薄纸质或坚纸质，长圆状椭圆形或倒披针形，长6～12cm，先端圆钝，基部楔形，全缘或微波状，无毛，无腺体，侧脉多而密，连网脉在两面均明显；叶柄长6～8mm。总状花序腋生或侧生，被微柔毛，长1～1.5cm；花4数，长3mm；花冠绿色至红色，裂片矩圆形或卵形，被腺体。果扁球形，径1～1.5cm，熟时红色，有纵肋及瘤点。

产通道、江永、江华、蓝山。生山谷林中或林缘。分布于福建、江西、广东、广西、香港、云南。果酸可食。

Evergreen climbing shrubs. Branchlets longitudinally angular. Leaves thinly papery or hard papery, oblong-elliptic or oblanceolate, 6-12cm long, apex obtuse, base cuneate, margin entire or sinuolate, glabrous, without glands, lateral veins numerous and dense, and with reticulate veins conspicuous on both surfaces; petioles 6-8mm long. Racemes axillary or lateral, puberulent, 1-1.5cm long; flowers 4-merous, 3mm long; corolla green to red, lobes oblong or ovate, with glands. Fruit oblate, 1-1.5cm in diam., red when mature, with longitudinal ribs and tubercles

Distributed in Hunan(Tongdao, Jiangyong, Jianghua, Lanshan), Fujian, Jiangxi, Guangdong, Guangxi, Hongkong and Yunnan. Grows in forests or forest margins in mountain valleys. Fruit is acid and edible.

攀援灌木。小枝纤细，排成2列，密被短腺毛。叶纸质，卵形或长卵形，长1～2.5cm，先端钝尖，基部截形至圆，两面密被腺体，上面中脉微凹，侧脉不明显；叶柄长1mm，被毛。花序腋生，近伞形或聚伞状，长5～10mm，被褐色微柔毛，花梗1～2mm；花5数，长2mm。果球形，径4～5mm，熟时暗红色，无毛，宿存花萼反卷。花期12月至翌年5月，果期5～7月。

产江永、宜章。生沟谷疏林或灌丛中。分布于福建、广东、广西、贵州、海南、西藏、云南、浙江南部。印度尼西亚、印度东北部、缅甸亦产。

Climbing shrubs. Branchlets slender, distichous, densely short glandular hairs. Leaves papery, ovate or long ovate, 1-2.5cm long, apex obtusely pointed, base truncate to rounded, densely glands on both surfaces, midvein slightly impressed adaxially, lateral veins inconspicuous; petioles 1mm long, pubescent. Panicles axillary, subumbellate or cymose, 5-10mm long, brown puberulent, pedicels 1-2mm long; flowers 5-merous, 2mm long. Fruit globose, 4-5mm in diam., dark red when mature, glabrous, persistent calyx reflexed. Fl. Dec to May of 2nd year, fr. May-Jul.

Distributed in Hunan(Jiangyong, Yizhang), Fujian, Guangdong, Guangxi, Guizhou, Hainan, Tibet, Yunnan and South Zhejiang, also in Indonesia, Northeast India, Burma. Grows in sparse forests or thickets in ravines.

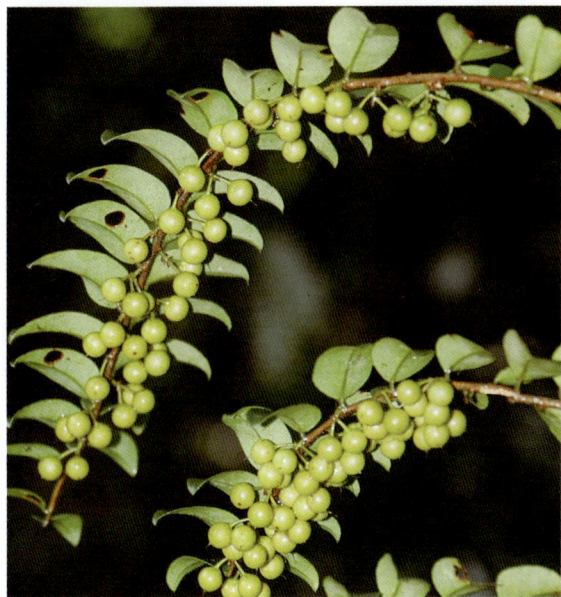

鲫鱼胆　　　*Maesa perlarius* (Lour.) Merr.

常绿灌木。叶纸质或坚纸质，宽椭圆状卵形至椭圆形，先端急尖或渐尖，长7～11cm，边缘上部具粗锯齿，下部全缘，幼叶密被毛，以后至少下面被长硬毛，亦有近无毛，侧脉7～8对，羽状直达齿端；叶柄长5～10mm，被长硬毛或疏柔毛。总状或圆锥花序腋生，长2～4cm，被疏柔毛或长硬毛；花钟形，白色。果球形或卵球形，径3mm，具纵条纹。花期3～4月，果期12月至翌年5月。

产通道、江永、永兴。生山地沟谷林下。分布于广东、广西、贵州、海南、四川、台湾、云南。越南亦产。

Evergreen shrubs. Leaves papery or hard papery, broadly elliptic-ovate to elliptic, apex acute or acuminate, 7-11cm long, margin coarsely serrate distally and entire toward base, young leaves densely hairy, later at least hirsute abaxially, sometimes subglabrous, lateral veins 7-8 pairs, pinnately straight to teeth; petioles 5-10mm long, hirsute or sparse pilose. Racemes or panicles, axillary, 2-4cm long, puberulous or hirsute; flowers bell-shaped, white. Fruit globose or ovoid, 3mm in diam., with longitudinal stripes. Fl. Mar-Apr, fr. Dec to May of 2nd year.

Distributed in Hunan(Tongdao, Jiangyong, Yongxing), Guangdong, Guangxi, Guizhou, Hainan, Sichuan, Taiwan and Yunnan, also in Vietnam. Grows in forests in mountain ravines.

常绿木质藤本。小枝圆柱形，幼时具棱。叶膜质，对生，卵形至卵状披针形，长5～12cm，先端渐尖，基部渐狭或近圆形，全缘，侧脉5～7对；叶柄长6～12mm。聚伞花序顶生或腋生；花小，淡黄色；花冠漏斗状，长1～1.5cm，内面具淡红色斑点，裂片5，卵形；雄蕊5，生花冠筒中部；花柱丝状，柱头4浅裂。蒴果卵形，熟时常黑色，基部具宿萼。花期5～11月，果期7月至翌年3月。

产通道。生山地灌丛或山坡疏林下。分布于福建、广东、香港、广西、贵州、海南、江西、台湾、云南、浙江南部。印度、缅甸、印度尼西亚亦产。全株有剧毒。

Evergreen woody vines. Branchlets cylindrical, angulate when young. Leaves membranous, opposite, ovate to ovate-lanceolate, 5-12cm long, apex acuminate, base attenuate or suborbicular, margin entire, lateral veins 5-7 pairs; petioles 6-12mm long. Cymes terminal or axillary; flowers small, pale yellow; corolla funnelform, 1-1.5cm long, inside with reddish spots, lobes 5, ovate; stamens 5, inserted at middle of corolla tube; style filiform, stigma 4-lobed. Capsule ovoid, usually black when mature, persistent calyx at base. Fl. May-Nov, fr. Jul to Mar of 2nd year. Distributed in Hunan(Tongdao), Fujian, Guangdong, Hongkong, Guangxi, Guizhou, Hainan, Jiangxi, Taiwan, Yunnan and South Zhejiang, also in India, Burma, Indonesia. Grows in mountain shrubs or sparse forests on mountain slopes. The whole plant is toxic.

海南链珠藤（串珠子） *Alyxia odorata* Wall. ex G. Don

攀援灌木，具乳汁。叶近革质，对生或3叶轮生，常集生枝顶，椭圆形至长圆形，长4～12cm，先端渐尖或急尖，基部宽圆，边缘微向外卷，侧脉多数，在叶面可见，在叶背不明显；叶柄长3～8mm。花黄色，多花集成聚伞花序，顶生和腋生。链珠状果实，具1～3近球形核果。花期8～10月，果期12月至翌年4月。

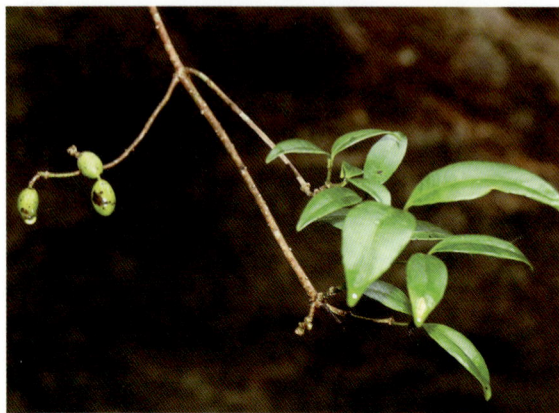

产通道、城步、江永、资兴、宜章。生海拔170～700m山地沟谷林中或林缘。分布于广东、广西、贵州、海南、四川、云南。缅甸、泰国亦产。

Climbing shrubs, with latex. Leaves subleathery, opposite or ternate, usually crowded at the top of branchlets, elliptic to oblong, 4-12cm long, apex acuminate or acute, base broadly rounded, margin slightly revolute, lateral veins numerous, visible adaxially and inconspicuous abaxially; petioles 3-8mm long. Flowers yellow, Cymes fascicled, terminal and axillary. Fruit moniliform, with 1-3 subglobose drupes. Fl. Aug-Oct, fr. Dec to Apr of 2nd year.

Distributed in Hunan(Tongdao, Chengbu, Jiangyong, Zixing, Yizhang), Guangdong, Guangxi, Guizhou, Hainan, Sichuan and Yunnan, also in Burma, Thailand. Grows in forests or forest margins in mountain ravines at alt. 170-700m.

链珠藤   *Alyxia sinensis* Champ. ex Benth.

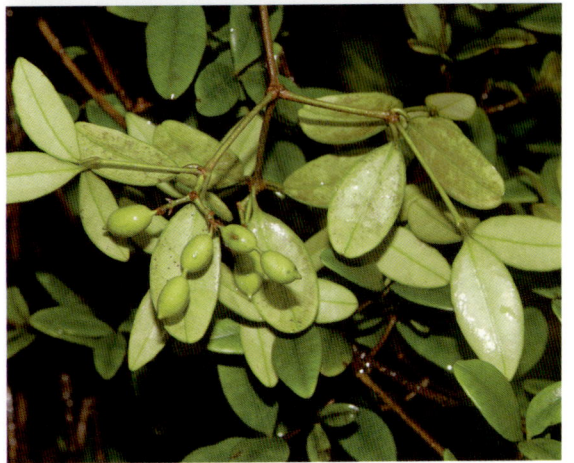

常绿木质藤本，具乳汁。小枝具4棱。叶革质，对生或3枚轮生，卵圆形或近圆形，长1.5～3.5cm，先端圆或微凹，两面无毛，边缘反卷，中脉上面下凹，侧脉不明显；叶柄长2mm。聚伞花序腋生或近顶生，花序长不及1.5cm；花小，长5～6mm；花冠白色，高脚碟状，花冠裂片5，向左覆盖。链珠状果实，具2～3卵形核果，核果长约1cm。花期5～9月，果期5～11月。

产通道、宜章。生山地沟谷林下。分布于福建、广东、广西、贵州、海南、江西、香港、台湾、浙江南部。

Evergreen woody vines, with latex. Branchlets 4-angled. Leaves leathery, opposite or ternate, ovate-orbicular or suborbicular, 1.5-3.5cm long, apex rounded or retuse, glabrous on both surfaces, margin revolute, midvein impressed adaxially, lateral veins inconspicuous; petioles 2mm long. Cymes axillary or subterminal, inflorescence less than 1.5cm long; flowers small, 5-6mm long; corolla white, salverform, corolla lobes 5, left arrow. Fruit moniliform, with 2-3 ovoid drupes, drupe ca.1cm long. Fl. May-Sep, fr. May-Nov.

Distributed in Hunan(Tongdao, Yizhang), Fujian, Guangdong, Guangxi, Guizhou, Hainan, Jiangxi, Hongkong, Taiwan and South Zhejiang. Grows in forests in mountain ravines.

大花帘子藤 *Pottsia grandiflora* Markgr.

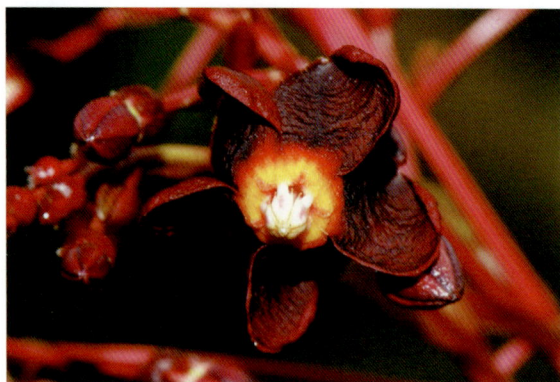

木质大藤本。枝条淡绿色，无毛，具乳汁。叶稍革质，卵圆形至椭圆状卵圆形，先端渐尖，基部钝至圆，长6.5～12.5cm，两面无毛，侧脉约6对；叶柄长1～2.2cm，叶柄间具钻状腺体。聚伞花序，顶生或腋生，长达20cm，无毛，花序梗长1～1.5cm；苞片和小苞片叶状；花蕾圆筒形，上部膨大，呈圆锥状，顶端钝；花冠紫红色或粉红色，长约1.3cm，无毛，花冠裂片倒卵形，反折；花柱长6mm，基部加厚，柱头圆锥状。蓇葖果2，线形，长达25cm，绿色，无毛；种子具白色种毛。花期4～8月，果期8～12月。

产江华、汝城、新宁、资兴。生海拔200～500m山地沟谷林中或灌丛。分布于福建、广东、广西、云南、浙江南部。

Woody large vines. Branches pale green, glabrous, with latex. Leaves somewhat leathery, ovate to elliptic-ovate, apex acuminate, base obtuse to rounded, 6.5-12.5cm long, glabrous on both surfaces, lateral veins ca.6 pairs; petioles 1-2.2cm long, with interpetiolar subulate glands. Cymes, terminal or axillary, up to 20cm long, glabrous, peduncle 1-1.5cm long; bracts and bracteoles leaflike; flower buds cylindrical, upper inflated, paniculate, apex obtuse; corolla purple-red or pink, ca.1.3 cm long, glabrous, corolla lobes obovate, reflexed; style 6 mm long, thickened near base; stigma conical. Follicles 2, linear, up to 25 cm long, green, glabrous; seeds with white hair. Fl. Apr-Aug, fr. Aug-Dec.

Distributed in Hunan(Jianghua, Rucheng, Xinning, Zixing), Fujian, Guangdong, Guangxi, Yunnan and South Zhejiang. Grows in forests or thickets in mountain ravines at alt. 200-500m.

毛杜仲藤　*Urceola huaitingii* (Chun et Tsang) D. J. Middl.

　　木质大藤本，具乳汁。全株密被灰色或红色短茸毛。叶对生，卵圆状或长圆状椭圆形，长2.5～8cm，先端锐尖，基部宽圆，两面被柔毛，下面脉上更密；叶柄长约5mm，被茸毛。伞状花序，近顶生及腋生；花冠黄色，坛状，外被微毛，裂片向左旋转。蓇葖双生或1个不发育，基部肿胀，种子具白色种毛。花期4～6月，果期7月至翌年6月。

　　产零陵、通道、江华、江永。生海拔200～600m山地林中。分布于广东、广西、贵州。

Woody large vines, with latex. Densely gray or red velutinous throughout. Leaves opposite, elliptic or oblong-elliptic, 2.5-8cm long, apex acute, base broadly rounded sides, puberulent on both surfaces, along veins more dense abaxially; petioles ca.5mm long, tomentose. Umbels, subterminal and axillary; corolla yellow, urn-shaped, puberulous, lobes left arrow. Follicles twins or 1agenesis, base swollen, seeds with white hair. Fl. Apr-Jun, fr. Jul to Jun of 2nd year.

Distributed in Hunan(Lingling, Tongdao, Jianghua, Jiangyong), Guangdong, Guangxi and Guizhou. Grows in mountain forests at alt. 200-600m.

酸叶胶藤 *Urceola rosea* (Hook. et Arn.) D. J. Middl.

木质大藤本，具乳汁，无皮孔。叶纸质，对生，宽椭圆形，长3～7cm，先端急尖，基部楔形，两面无毛，下面被白粉，侧脉4～6对。聚伞花序，花序梗疏被白粉和短柔毛；花萼5深裂，裂片卵形；花冠粉红色，坛状，花冠裂片5，向右覆盖；雄蕊5；花盘环状。蓇葖果双生，叉开成一直线，具明显斑点，长达15cm；种子具白色毛。花期4～12月，果期7月至翌年1月。

产江华、宜章。生海拔500m以下山地沟谷林中。分布于福建、广东、广西、贵州、海南、香港、四川、台湾、云南。印度尼西亚、泰国、越南亦产。

Wooden large vines, with latex, lenticels absent. Leaves papery, opposite, broadly elliptic, 3-7cm long, apex acute, base cuneate, glabrous on both surfaces, pruinose abaxially, lateral veins 4-6 pairs. Cymes, peduncle sparsely pruinose and pubescent; calyx deeply 5-lobed, lobes ovate; corolla pink, urn-shaped, corolla lobes 5, right arrow; stamens 5; disk annular. Follicles twins, opened in a straight line, with conspicuous spots, up to 15cm long; seeds with white hair. Fl. Apr-Dec, fr. Jul-Jan of following year.

Distributed in Hunan(Jianghua, Yizhang), Fujian, Guangdong, Guangxi, Guizhou, Hainan, Hongkong, Sichuan, Taiwan and Yunnan, also in Indonesia, Thailand, Vietnam. Grows in forests in mountain ravines below alt. 500m.

常绿灌木。叶纸质或革质，对生，长圆形或椭圆状长圆形，长5～24cm，先端渐尖或短尖，基部楔形，全缘，侧脉5～15对，连中脉在上面下凹，脉腋内常有束毛；叶柄长0.7～5cm，近无毛；托叶膜质，短鞘状。聚伞花序常顶生，具多花，总花梗常极短；花梗长1～2.5mm；萼管杯状，檐部扩大；花冠白色，喉部被白色长柔毛，花冠裂片近三角形，开放时反折。核果球形或宽椭圆形，有纵棱，熟时红色。花果期全年。

产宜章。生石灰岩山地林下。分布于福建、台湾、广东、香港、海南、广西、贵州、云南。日本、南亚、东南亚亦产。

Evergreen shrubs. Leaves papery or leathery, opposite, oblong or elliptic-oblong, 5-24cm long, apex acuminate or short acuminate, base cuneate, margin entire, lateral veins 5-15 pairs, and with midrib impressed adaxially, usually tufted hairs in vein axils; petioles 0.7-5cm long, subglabrous; stipules membranous, like short sheath. Cymes usually terminal, with many flowers, peduncle usually very short; pedicels 1-2.5mm long; calyx tube cupular, brim expanded; corolla white, throat white villous, corolla lobes subtriangular, reflexed when opening. Drupe globose or broadly elliptic, with longitudinal ribs, red when mature. Fl. and fr. year-round.

Distributed in Hunan(Yizhang), Fujian, Taiwan, Guangdong, Hongkong, Hainan, Guangxi, Guizhou and Yunnan, also in Japan, South Asia, Southeast Asia. Grows in forests in limestone mountain.

常绿攀援藤本，常以气生根攀附于树干或岩石上。小枝稍扁，老枝柱状，近木质，攀附枝有一列短而密的气根。叶厚纸质，对生，椭圆形至卵形，或倒卵形至倒披针形，长2～6cm，先端钝尖，基部楔形或稍圆，边缘反卷，侧脉4～10对，不明显；叶柄长3～5mm。聚伞花序顶生，总花梗长达3cm；花小，白色。核果近球形，径4～7mm，具纵棱，常呈白色。花期4～6月，果期全年。

产汝城。生低山灌丛或林中。分布于浙江南部、福建、台湾、广东、香港、海南、广西。越南、日本亦产。湖南省新记录种。

Evergreen climbing vines, usually cling to the trunk or on rocks by aerial roots. Branchlets slightly compressed, old branches columnar, nearly woody, climbing branches with a row of short and dense aerial root. Leaves thickly papery, opposite, oblong to ovate, or obovate to oblanceolate, 2-6cm long, apex obtusely pointed, base cuneate or slightly rounded, margin revolute, lateral veins 4-10 pairs, inconspicuous; petioles 3-5mm long. Cymes terminal, peduncle up to 3cm long; flowers small, white. Drupe subglobose, 4-7mm in diam., with longitudinally angular, usually white. Fl. Apr-Jun, fr. year-round.

Distributed in Hunan(Rucheng), South Zhejiang, Fujian, Taiwan, Guangdong, Hongkong, Hainan and Guangxi, also in Vietnam, Japan. Grows in thickets or forests in low mountain. A new recorded species in Hunan province.

伞花螺序草 *Spiradiclis umbelliformis* H. S. Lo

匍匐草本。茎密被棕红色长柔毛，下部节上生根。叶卵状圆形，长1.5～4cm，先端钝或圆，基部心形，全缘，被缘毛，侧脉4～6对，下面苍白色，沿中脉及侧脉被疏柔毛；叶柄长1～3cm，被棕红色长柔毛，托叶2深裂，裂片线形。聚散花序，伞形或近头状，顶生，具4～10花，总梗长2～7cm；花近无梗，白色或微红紫，2型，花柱异长，长柱花近漏斗状，花柱长约1.4cm；短柱花花柱长2mm。蒴果近球形，具5直棱。花期4月。

产道县。生于林下岩石壁中。分布于广西、广东。湖南新记录属、种。

Creeping herbs. Stem densely reddish brown villous, rooting at lower nodes. Leaves ovate-orbicular, 1.5-4cm long, apex obtuse or rounded, base cordate, margin entire, ciliate, lateral veins 4-6 pairs, whitish abaxially, along midrib and lateral veins puberulous; petioles 1-3cm long, reddish-brown villous; stipules deeply 2-lobed, lobes linear. Inflorescences cymose, umbelliform or subcapitate, terminal, with 4-10 flowers, peduncle 2-7cm long; flowers subsessile, white or pale purplish red, dimorphic, style heterostylous, long style flowers: nearly funnelform, style ca.1.4cm long; short style flowers: style 2mm long. Capsule subglobose, with 5 straight ribs. Fl. Apr.

Distributed in Hunan(Daoxian), Guangxi and Guangdong. Grows on rocks in forests. New record genus, species in Hunan province.

白花丹 *Plumbago zeylanica* L.

常绿半灌木。小枝平展，无毛。叶纸质，卵形至长圆状卵形，长3～10cm，先端急尖至渐尖，基部宽楔形，无毛，叶柄基部有时耳形。穗状花序，顶生和腋生，长5～25cm，花序轴具腺体；花萼绿色，长约1cm，顶端5裂，具5棱，具腺毛；花冠高脚碟状，白色，筒长约2cm，顶端5裂；雄蕊5。蒴果淡黄褐色，长圆形；种子红棕色。花期10月至翌年3月，果期12月至翌年4月。

产江华、江永。生阴湿处或荒坡灌丛。分布于福建、广东、广西、贵州、海南、四川、台湾、云南。世界其他热带地区亦产。

Evergreen subshrubs. Branches spreading, glabrous. Leaves papery, ovate to oblong-ovate, 3-10cm long, apex acute to acuminate, base broadly cuneate, glabrous; petioles base sometimes auriculate. Spikes, terminal and axillary, 5-25cm long, inflorescences rachis with glands; calyx green, ca.1cm long, apically 5-lobed, 5-angled, glandular hairs; corolla salverform, white, tube ca.2cm long, apically 5-lobed; stamens 5. Capsule pale yellow-brown, oblong; seeds red-brown. Fl. Oct-Mar of 2nd year, fr. Dec to Apr of 2nd year.

Distributed in Hunan(Jianghua, Jiangyong), Fujian, Guangdong, Guangxi, Guizhou, Hainan, Sichuan, Taiwan and Yunnan, also in other tropical regions of the world. Grows in moist places or barren shrubs.

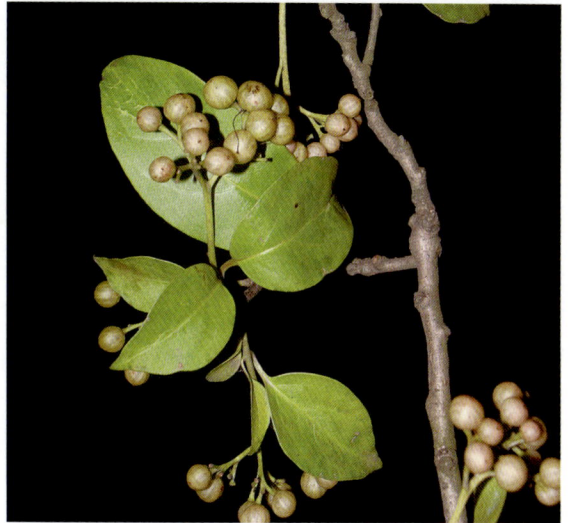

常绿乔木。小枝无毛。叶薄革质，长圆形或椭圆形，长7～16cm，全缘，无毛，侧脉4～7对，网脉不明显，干后呈黑色；叶柄长1～2.5cm。圆锥花序伞房状，长4～8cm；花萼长约2mm，无毛，5浅裂；花冠白色，长12mm，裂片5，卵形，冠筒长7～10mm；雄蕊5，伸出花冠之外；花柱2裂。核果近球形，暗紫色，径7～10mm。花期4月，果期6～7月。

产绥宁、通道、道县、江华、城步、洞口、宜章、永兴、资兴、炎陵、汝城。生低山阔叶林中。分布于福建、广东、广西、台湾、江西南部。越南北部亦产。

Evergreen trees. Branchlets glabrous. Leaves thinly leathery, oblong or elliptic, 7-16cm long, margin entire, glabrous, lateral veins 4-7 pairs, reticulate veins inconspicuous, after dry black; petioles 1-2.5cm long. Panicles like corymbose, 4-8cm long; calyx ca.2mm long, glabrous, 5-lobed; corolla white, 12mm long, lobes 5, ovate, corolla tube 7-10mm long; stamens 5, stretched corolla outside; style 2-cleft. Drupe subglobose, dark purple, 7-10mm in diam.. Fl. Apr, fr. Jun-Jul.

Distributed in Hunan(Suining, Tongdao, Daoxian, Jianghua, Chengbu, Dongkou, Yizhang, Yongxing, Zixing, Yanling, Rucheng), Fujian, Guangdong, Guangxi, Taiwan and South Jiangxi, also in North Vietnam. Grows in broad-leaved forests in low mountain.

毛麝香　　*Adenosma glutinosum* (Linn.) Druce

多年生直立草本。全株密被腺毛和长柔毛。茎基部木质化，高达1m。叶对生，卵状披针形至宽卵形，长2～8cm，边缘具不整齐锯齿，叶背面、苞片、小苞片和萼片均被黄色透明腺体；具短柄或近无柄。花腋生，单生或总状花序顶生；萼片5枚，果期稍增大宿存；花冠蓝色或紫红色，长1～2.5cm，上唇直立，圆卵形、截形或微凹，下唇3裂。蒴果卵形，先端具喙，4瓣裂。花果期7～10月。

产江华、江永、汝城。生低海拔沟谷边或湿润山坡疏林下。分布于福建西南部、广西、海南、江西南部、云南。印度至马来西亚、大洋洲亦产。叶含芳香油，亦可药用。湖南省新记录属、种。

Perennial erect herbs. Densely glandular hairs and villous throughout. Stem basally lignified, up to 1m tall. Leaves opposite, ovate-lanceolate to broadly ovate, 2-8cm long, margin irregularly serrate, leaf abaxially, bracts, bracteoles, and sepals with transparent yellow glands; short petiolate or subsessile. Flowers axillary and solitary or racemes terminal; sepals 5, slightly increased and persistent when fruit; corolla blue or purplish-red, 1-2.5cm long, upper lip erect, ovate, truncate or retuse, lower lip 3-cleft. Capsule ovoid, apex beaked, 4-valved. Fl. and fr. Jul-Oct.

Distributed in Hunan(Jianghua, Jiangyong, Rucheng), Southwest Fujian, Guangxi, Hainan, South Jiangxi and Yunnan, also in India to Malaysia, Oceania. Grows in ravine sides or sparse forests on wet mountain slopes at low altitude. Leaves contain essential oil, and also used medicinally. New recorded genus, species in Hunan province.

glabrescent. Leaves opposite, ovate-orbicular, rarely ovate-oblong, 3-11cm long, apex acute to caudate-acute, base broadly cuneate or subcordate, margin entire or sparse serrate, after dry black; petioles up to 8mm long. Flowers solitary and axillary, sometimes paired; calyx campanulate, up to 1.5cm long, inside sericeous, 5-lobed to 1/2 of calyx length; calyx teeth narrowly triangular-ovate, apex acuminate and apiculate; corolla yellow, ca.2.5cm long, upper lip 2-cleft, lobes obliquely ovate. Capsule small, shorter than sepals. Fl. Jun-Nov, fr. Dec to Jan of 2nd year.

Distributed in Hunan(Daoxian, Ningyuan, Lanshan, Lingwu, Yizhang, Rucheng), Guangdong and Guangxi. Grows in shrubs in mountain land.

直立灌木。全体密被灰棕色星状茸毛，枝、叶上面渐变无毛。叶对生，卵圆形，稀卵状长圆形，长3～11cm，先端锐尖或尾状锐尖，基部宽楔形或近心形，全缘或具疏齿，干后呈黑色；叶柄长达8mm。花单生叶腋，偶2花同生；萼钟形，长达1.5cm，内面被绢毛，5裂至1/2处；萼齿狭三角状卵形，先端渐狭成长尖头；花冠黄色，长约2.5cm，上唇2裂，裂片歪卵形。蒴果小，短于萼片。花期6～11月，果期12月至翌年1月。

产道县、宁远、蓝山、临武、宜章、汝城。生山地灌丛。分布于广东、广西。

Erect shrubs. Densely gray-brown stellate tomentose throughout. Branches, leaf adaxially

落叶灌木。叶对生，2(3)回奇数羽状复叶，小叶椭圆形或卵形，长3～7cm，先端长尾状，基部宽楔形，全缘，侧脉5～6对，侧生小叶柄短于5mm，顶生小叶柄长1～2cm。圆锥花序顶生，具很长的苞片及小苞片；花萼卵形，萼齿5；花冠黄白色，筒状，裂片5，圆形，具皱纹，雄蕊4。蒴果圆柱形，下垂，常扭曲，长可达80cm，径约1cm，2纵裂。花期5～9月，果期10～12月。

产道县、江永。生海拔200-400m石灰岩山丘、村边林。分布于广东、广西、贵州、台湾、云南。不丹亦产。株形雅致，复叶硕大，宜作园林观赏树种。

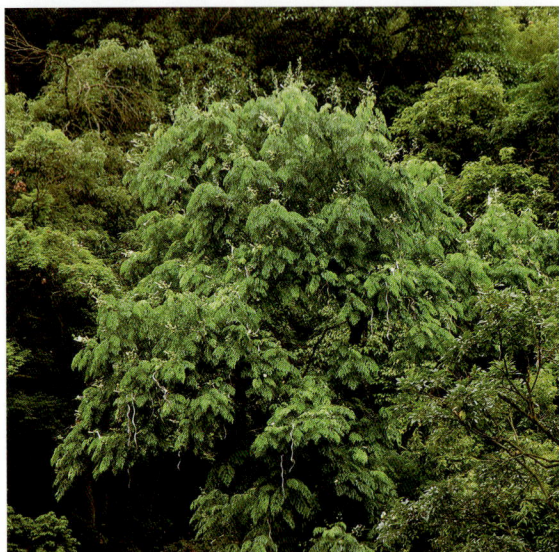

Deciduous shrubs. Leaves opposite, 2 (3) odd pinnae, leaflets elliptic or ovate, 3-7cm long, apex long caudate, base broadly cuneate, margin entire, lateral veins 5-6 pairs, lateral petiolules less than 5 mm, terminal petiolules 1-2cm long. Inflorescences paniculate terminal, with long bracts and bracteoles; calyx ovate, Calyx teeth 5; corolla white, tubular, lobes 5, rounded, rugose; stamens 4. Capsule terete, nodding, usually distorted, up to 80cm long, ca.1cm in diam. 2 longitudinal split. Fl. May-Sep, fr. Oct-Dec.

Distributed in Hunan(Daoxian, Jiangyong), Guangdong, Guangxi, Guizhou, Taiwan and Yunnan, also in Bhutan. Grows in limestone hills, and in forests of village edge at alt. 200-400m. It is an excellent ornamental tree species for its elegant tree-shape and large compound leaves.

钟花草　*Codonacanthus pauciflorus* (Ness) Ness

纤细草本，多分枝或不分枝。叶薄纸质，椭圆状卵形或狭披针形，长3～8cm，先端急尖或渐尖，全缘或有时波状，两面被微柔毛，侧脉纤细，5～7对；叶柄长5～10mm。花序总状，细长，稀疏；花梗长1～2mm；花萼裂片5，三角状披针形；花冠白色，钟状，下部偏斜，裂片5，几相等。蒴果无毛，具4种子。花果期8月至翌年4月。

产江永。生海拔225m潮湿山谷。分布于广东、香港、广西、海南、台湾、福建、贵州、云南、澳门。越南至印度亦产。湖南省新记录属、种。

Slender herbs, much branched or unbranched. Leaves thinly papery, elliptic-ovate or narrowly lanceolate, 3-8cm long, apex acute or acuminate, margin entire or sometimes undulate, puberulent on both surfaces, lateral veins slender, 5-7 pairs; petioles 5-10mm long. Inflorescence racemose, slender, sparsely; pedicels 1-2mm long; calyx lobes 5, triangular-lanceolate; corolla white, campanulate, lower oblique, lobes 5, subequal in length. Capsule glabrous, with 4 seeds. Fl. and fr. Aug-Apr of 2nd year.

Distributed in Hunan(Jiangyong), Guangdong, Hongkong, Guangxi, Hainan, Taiwan, Fujian, Guizhou, Yunnan and Macao, also in Vietnam to India. Grows in wet valleys at alt. 225m. New recorded genus, species in Hunan province.

弯花叉柱花 *Staurogyne chapaensis* R. Ben.

草本，高达10cm。茎缩短。叶莲座状基生，疏被棕色长柔毛，叶片卵形或长圆形，先端钝，基部心形，上面被稀疏长柔毛，下面苍白色，几无毛，侧脉7~9对，全缘或波状。总状花序顶生，具多花；苞片互生，倒卵形或线状匙形；花梗长3mm，被长柔毛；花萼不等5裂；花冠淡蓝紫色，冠檐5裂，裂片圆形，近相等，能育雄蕊4，2长2短。花期3~5月。

产道县、汝城。生山地密林下。分布于广东、广西及云南东南部。越南北部亦产。湖南省新记录种。

Herbs, up to 10cm tall. Stems short. Leaves in a basal rosette, sparsely brown villous, leaf blade ovate or oblong, apex obtuse, base cordate, sparsely villous adaxially, pale abaxially, glabrescent, lateral veins 7-9 pairs, margin entire or undulate. Racemes terminal, with many flowers; bracts alternate, obovate or linear spatulate; pedicels 3mm long, villous; calyx unequal 5-cleft; corolla light bluish purple, limb 5-cleft, lobes orbicular, subequal in length, fertile stamens 4, 2 long, 2 short. Fl. Mar-May.

Distributed in Hunan(Daoxian, Rucheng), Guangdong, Guangxi and Southeast Yunnan, also in North Vietnam. Grows in dense forests in mountain land. A new recorded species in Hunan province.

落叶灌木。小枝紫褐色，4棱形。叶披针形至狭椭圆形，长14～23cm，先端锐尖，边缘具不明显的细齿或近全缘，上面仅脉上被毛，下面无毛，被细小黄色腺体；叶柄长1～2cm。聚伞花序腋生，花小而密集；花序梗1.5～3cm；花萼无毛，萼齿不明显；花冠紫红色，无毛。果实扁球形，无毛，被红色腺体。花期7～9月，果期10～12月。

产通道、江永。生山坡或旷野。分布于福建、广东、广西、海南、江西、四川、台湾。越南、菲律宾亦产。

Deciduous shrubs. Branchlets purplish brown, 4-angled. Leaves lanceolate to narrowly elliptic, 14-23cm long, apex acute, margin obscurely serrulate or subentire, only veins hairy adaxially, glabrous abaxially, with tiny yellow glands; petioles 1-2cm long. Cymes, axillary, flowers small and dense; peduncle 1.5-3cm long; calyx glabrous, calyx teeth inconspicuous; corolla purple-red, glabrous. Fruit oblate, glabrous, with red glands. Fl. Jul-Sep, fr. Oct-Dec.

Distributed in Hunan(Tongdao, Jiangyong), Fujian, Guangdong, Guangxi, Hainan, Jiangxi, Sichuan and Taiwan, also in Vietnam, Philippines. Grows on mountain slopes or wilderness.

白花灯笼　*Clerodendrum fortunatum* L.

落叶灌木。嫩枝暗棕褐色，被短柔毛。叶一般为椭圆形或卵状披针形，长5～16cm，先端渐尖，基部宽楔形，全缘或波状。聚伞花序腋生，短于叶，具3～9花；花萼紫红色，5深裂，裂片宽卵形；花冠淡红色或白色稍带紫。核果近球形，熟时深蓝色，藏于萼内。花果期6～11月。

产桂东、通道、宜章。生山丘沟谷或旷野。分布于福建、广东、广西、江西、海南。

Deciduous shrubs. Young branchlets dark brown, pubescent. Leaves usually elliptic or ovate-lanceolate, 5-16cm long, apex acuminate, base broadly cuneate, margin entire or undulate. Cymes, axillary, shorter than leaf, with 3-9 flowers; calyx purplish-red, deeply 5-lobed, lobes broadly ovate; corolla pale red or white with purple. Drupe subglobose, deep blue when mature, included within calyx. Fl. and fr. Jun-Nov.

Distributed in Hunan(Guidong, Tongdao, Yizhang), Fujian, Guangdong, Guangxi, Jiangxi and Hainan. Grows in mountain ravines or desert hills.

大苞鸭跖草　*Commelina paludosa* Bl.

多年生粗壮草本。茎常直立，不分枝，或少有上部分枝，无毛或疏生短硬毛。叶披针形至卵状披针形，先端长渐尖；叶鞘密生棕色长睫毛。总苞片常4～10个集生茎顶，呈扁漏斗状；蝎尾状聚伞花序具数花；花瓣3枚，蓝色，匙形或倒卵状圆形。蒴果卵球形，三棱形，3瓣裂。花期8～10月，果期10月至翌年4月。

产通道、新宁、宜章。生海拔500m以下山谷湿地、溪边。分布于福建、广东、广西、贵州、江西南部、四川、台湾、西藏、云南。印度至马来西亚亦产。

Perennial stout herbs. Stem often erect, simple or sometimes branched distally, glabrous or sparsely hispidulous. Leaves lanceolate to ovate-lanceolate, apex long acuminate; leaf sheaths densely brown long eyelashes. Involucral bracts 4-10 crowded at the top of stem, compressed funnelform; Cincinnus with several flowers; petals 3, blue, spatulate or obovate-oblong. Capsule ovoid-globose, trigonous, 3-valved. Fl. Aug-Oct, fr. Oct to Apr of 2nd year.

Distributed in Hunan(Tongdao, Xinning, Yizhang), Fujian, Guangdong, Guangxi, Guizhou, South Jiangxi, Sichuan, Taiwan, Tibet and Yunnan, also in India to Malaysia. Grows in valley wetland and streamsides at alt. 500m.

多年生草本。根状茎极长，全体被多细胞柔毛，叶鞘及花序各部分较密。茎高15～50cm，不分枝。叶长椭圆形至披针形，长4～10cm，上面有鳞片状突起；叶无柄或具带翅短柄。聚伞花序组成扫帚状圆锥花序；花梗短；苞片鳞片状；萼片浅舟状；花瓣蓝色或紫色，倒卵形。蒴果卵圆形，压扁。花果期7～11月。

产江华、通道、宜章、汝城。生低海拔山地林下或溪沟边。分布于福建、广东、广西、海南、江西、四川、西藏、云南、浙江南部。印度、东南亚、大洋洲热带地区亦产。

Perennial herbs. Rhizomes extremely long, multicellular hairs throughout, rather dense on leaf sheaths and inflorescences. Stem 15-50cm tall, unbranched. Leaves elliptic to lanceolate, 4-10cm long, scalelike protuberances adaxially; leaves sessile or with short, winged petiole. Cymes arranged in broomlike panicle; pedicels short; bracts scalelike; sepals shallowly boat-shaped; petals blue or purple, obovate. Capsule ovate-orbicular, compressed. Fl. and fr. Jul-Nov.

Distributed in Hunan (Jianghua, Yizhang, Tongdao, Rucheng), Fujian, Guangdong, Guangxi, Hainan, Jiangxi, Sichuan, Tibet, Yunnan and South Zhejiang, also in India, Southeast Asia, tropical area of Oceania. Grows in forests or streamsides in mountain at low altitude.

野 蕉   *Musa balbisiana* Colla

树状草本。假茎丛生，黄绿色。叶片卵状长圆形，长达3m，基部耳形，两侧不对称，下面微被白粉。花序下垂，长达2.5m；雌花苞片脱落，中性花及雄花苞片宿存，苞片卵形至披针形，外面暗紫红色，被白粉，内面紫红色，开放后反卷。果序下垂，约具8果丛，每果丛排成2行。浆果倒卵形，灰绿色，熟时具明显棱角。花期夏秋。

产通道、江永。生海拔500m以下山谷湿地。分布于广东、广西、海南、西藏、云南。亚洲南部及东南部亦产。假茎粉碎可作饲料。

Arborescent herbs. Pseudostems clumped, yellow-green. Leaf blade ovate-oblong, up to 3m long, base auriculate, asymmetry, slightly pruinose abaxially. Inflorescence pendulous, up to 2.5m long; bracts of female flowers deciduous, bracts of neutral and male flowers persistent, bracts ovate to lanceolate, outside dark purple, pruinose, inside purple red, reflexed after flowering. Infructescence pendulous, with ca.8 clusters, each cluster arranged in 2 rows. Berry obovoid, gray-green, distinctly angled at maturity. Fl. Summer and Autumn.

Distributed in Hunan(Tongdao, Jiangyong), Guangdong, Guangxi, Hainan, Tibet and Yunnan, also in South Asia and Southeast Asia. Grows in valley wetland below alt. 500m. The crushed pseudostems can be used as feed.

花叶山姜　*Alpinia pumila* Hook. f.

多年生草本。无地上茎，根茎平卧。叶2～3片，直立；叶片椭圆形、长圆形或长圆状披针形，上面绿色，沿脉具深绿色条纹，下面浅绿色，两面无毛；叶柄长约2cm；叶鞘红褐色。总状花序自叶鞘间抽出；花成对生长；萼片紫红色，被短柔毛；花冠白色；唇瓣卵形，顶端短2裂，反折。果球形，顶端具宿存花被片。花期4～6月，果期6～11月。

产江华、宜章、洞口。生山地阴湿林下。分布于江西南部、广东、广西、云南。

Perennial herbs. Without the bine on the ground, rhizomes procumbent. Leaves 2 or 3, erect; leaf blade elliptic, oblong or oblong-lanceolate, green abaxially, with deep green stripes along veins, light green abaxially, glabrous on both surfaces; petioles ca.2cm long; leaf sheaths reddish-brown. Racemes from between leaf sheaths; flowers paired; sepals purplish-red, pubescent; corolla white; labellum ovate, apex short 2-cleft, reflexed. Fruit globose, apex with persistent perianth. Fl. Apr-Jun, fr. Jun-Nov.

Distributed in Hunan(Jianghua, Yizhang, Dongkou), South Jiangxi, Guangdong, Guangxi and Yunnan. Grows in damp forests in mountain land.

多年生草本。叶片长圆状披针形或披针形，长20～40cm，下面被短柔毛；无柄；叶舌薄膜质，长2～3cm。穗状花序常顶生，长10～20cm；苞片卵圆形，覆瓦状排列，每一苞片内具2～3花；花白色，花萼管长约4cm；花冠管纤细，长8cm，裂片披针形，中部1个匙形；唇瓣倒心形，顶端2裂。花期8～12月。

产新晃。生海拔700m以下山地林下。分布于广东、广西、四川、台湾、云南。印度、越南、马来西亚至澳大利亚亦产。花芳香，庭园常栽培。

Perennial herbs. Leaf blade oblong-lanceolate or lanceolate, 20-40cm long, pubescent abaxially; leaves sessile; ligule membranous, 2-3cm long. Spikes usually terminal, 10-20cm long; bracts ovate-orbicular, imbricate arrangement, each bract with 2-3 flowers; flowers white, calyx tube ca.4cm long; corolla tube slender, 8cm long, lobes lanceolate, central one spatulate; labellum obcordate, apex 2-cleft. Fl. Aug-Dec.

Distributed in Hunan(Xinhuang), Guangdong, Guangxi, Sichuan, Taiwan and Yunnan, also in India, Vietnam, Malaysia to Australia. Grows in mountain forests below alt. 700m. The flowers are fragrant, usually cultivated as an ornamental plant.

海 芋　　*Alocasia macrorrhiza* (L.) Schott

大型常绿草本。茎粗壮，圆柱形，有节，常生不定芽条。叶多数，叶柄粗大，绿色或淡紫色；叶片亚革质，草绿色，箭状卵形，边缘波状。花序柄2～3枚丛生，圆柱形；佛焰苞管部席卷呈长圆状卵形或卵形，檐部舟状，长圆形，略下弯。浆果卵状，红色。花果期4～8月。

产江永、通道、宜章。生山谷林下、水沟边。分布于台湾、福建、江西、广东、广西、贵州、云南。南亚及东南亚亦产。叶硕大浓绿，宜作园林观赏。

Large evergreen herbs. Stem stout, terete, noded, usually with adventitious buds. Leaves numerous, petioles thick and big, green or light purple; leaves subleathery, lawngreen, arrow-shaped ovate, margin undulate. Peduncle 2-3-clustered, terete; spathe curled at tube, oblong-ovate or ovate, limb navicular, oblong, slightly recurved. Berry ovate, red. Fl. and fr. Apr-Aug.

Distributed in Hunan(Jiangyong, Tongdao, Yizhang), Taiwan, Fujian, Jiangxi, Guangdong, Guangxi, Guizhou and Yunnan, also in South Asia and Southeast Asia. Grows in mountain valley forests, and streamsides. It can be used as an ornamental plant for its large and dense green leaves.

多年生常绿草本。根茎倒圆锥形。叶丛生，叶柄淡绿色，被白粉，长可达1.5m，下部1/2鞘状，闭合；叶片卵状心形，先端短渐尖，基部心形或盾形，边缘波状。花序柄近圆柱形，常5～8枚生于叶腋。佛焰苞管部绿色，椭圆状，席卷；檐部粉白色，长圆形，基部兜状，直立。肉穗花序长9～20cm，雌花序圆锥状，奶黄色；附属器极短小，锥状。浆果圆柱形，种子多数，纺锤形。花期4～6月，果期9月。

产通道、宁远、保靖、双牌。生沟谷林下湿地。分布于广东、广西、云南、福建、江西。马来半岛、中南半岛亦产。

Perennial evergreen herbs. Root obconic. Leaves clustered, petioles light green, up to 1.5m long, proximal half sheathing, closed; leaf blade ovate-cordate, apex shortly acuminate, base cordate or peltate, margin undulate. Peduncle subterete, usually 5-8 arising from leaf axil. Spathe tube green, elliptic, curled; limb white, oblong, base cucullate, erect. Spadix 9-20cm long, female inflorescences paniculate, milk yellow; appendix very short, conical. Berry terete, seeds numerous, fusiform. Fl. Apr-Jun fr.Sep.

Distributed in Hunan(Tongdao, Ningyuan, Baojing, Shuangpai), Guangdong, Guangxi, Yunnan, Fujian and Jiangxi, also in Malay Peninsula, Indo-China Peninsula. Grows in wetland in ravine forests.

攀援藤本，茎丛生。叶羽状全裂，裂片多数，2列，互生或近对生，条形，边缘、主脉被刚毛状刺；叶轴三棱形，被稀疏长或短的锐钩刺；叶鞘具纤鞭，无刺或少刺。肉穗花序长，具爪状纤鞭；雄花序具2回羽状分枝；雌花序仅有1回羽状分枝。果椭圆形，黄白色，长1～1.2cm，顶端喙状。花果期5～6月。

产汝城。生山地密林中。分布于福建、广东、香港、广西、海南、贵州、云南。茎可编织各种藤器，果可食。湖南省新记录种。

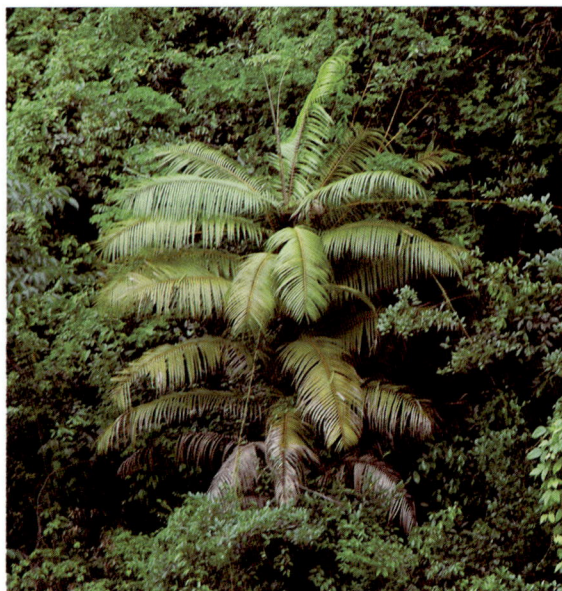

Climbing vines, stems clustered. Leaves pinnatisect, lobes numerous, distichous, alternate or subopposite, strip, setiform spines on margin and main veins; leaf rachises prismatoidal, sparsely long or short sharp hooked spines; leaf sheaths with flagella, no spines or less spines. Spadix long, clawed flagellate; male inflorescence 2 pinnately branched; female inflorescences only 1 pinnately branched. Fruit elliptic, yellow-white, 1-1.2cm long, apex beak-like. Fl. and fr. May-Jun.

Distributed in Hunan(Rucheng), Fujian, Guangdong, Hongkong, Guangxi, Hainan, Guizhou and Yunnan. Grows in dense forests in mountain land. The stems are used for weaving all kinds of rattan furniture, and the fruits are edible. A new recorded species in Hunan province.

直立灌木，茎丛生。叶羽状全裂，羽片多数，狭披针形或披针形，两面黄绿色；叶轴背面具爪状刺；叶鞘被棕褐鳞秕和褐色长刺；托叶鞘紫色，无毛。雄花序圆锥状，直立或拱垂，3回分枝；雌花序长45～60cm，直立，具2回分枝。果卵形或椭圆形，长15mm，具短圆锥状的喙。花期6～7月，果期9～10月。

产江永、道县。生沟谷溪边。分布于广东、广西、江西、福建、浙江南部。

Erect shrubs, stems clustered. Leaves pinnatisect, pinnae numerous, narrowly lanceolate or lanceolate, yellow green on both surfaces; leaf rachises abaxially clawed spines; leaf sheaths with brown scales and brown long spines; ocrea purplish, glabrous. Male inflorescences paniculate, erect or arching down, three branches; female inflorescences 45-60cm long, erect, with two branches. Fruit ovoid or oblong, 15mm long, with short conical beak. Fl. Jun-Jul, fr. Sep-Oct.

Distributed in Hunan(Jiangyong, Daoxian), Guangdong, Guangxi, Jiangxi, Fujian and South Zhejiang. Grows in ravine streamsides.

多年生草本。根茎肥大，具多数须根。叶基生，叶片椭圆状披针形，长10～22cm，先端渐尖，基部下延，全缘，叶脉在上面凹，在下面突起；叶柄长7～11cm。花葶自叶丛中抽出，长6～13cm；伞形花序顶生，具8～15花；总苞4枚，卵形或三角状卵形，外面2枚较大，内面2枚稍小；苞片线形，长达7cm；花被钟状，裂片6，外面淡绿色，内面淡紫色；雄蕊6，花药淡紫色。蒴果近倒卵形，3瓣裂。花期5～6月，果期7～8月。

产东安、江华、新宁、武冈、宜章。生海拔200～600m山地溪边、林下、田边。分布于广东、广西、贵州、江西南部、云南。泰国、越南、老挝亦产。

Perennial herbs. Rhizomes hypertrophy, with numerous fibrous roots. Leaves basal, leaf blade elliptic-lanceolate, 10-22cm long, apex acuminate, base decurrent on petiole, margin entire, veins impressed adaxially, prominent abaxially; petioles 7-11cm long. Scapes from basal leaves, 6-13cm long; umbels terminal, with 8-15 flowers; involucral bracts 4, ovate or triangular-ovate, outside 2 larger, inside 2 slightly small; bracts linear, up to 7cm long; perianth campanulate, lobes 6, outside pale green, inside pale purple; stamens 6, anthers pale purple. Capsule subobovoid, 3-valved. Fl. May-Jun, fr. Jul-Agu.

Distributed in Hunan(Dongan, Jianghua, Xinning, Wugang, Yizhang), Guangdong, Guangxi, Guizhou, South Jiangxi and Yunnan, also in Thailand, Vietnam, Laos. Grows in streamsides, forests, cropland edges in mountain land at alt. 200-600m.

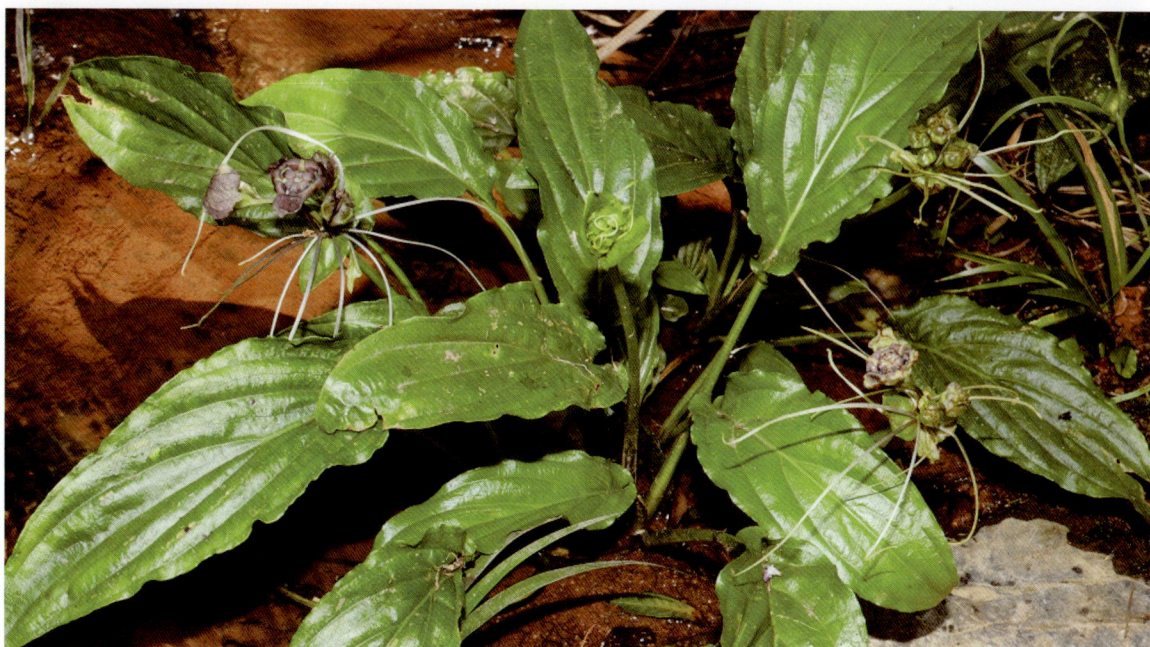

多年生草本。根状茎粗壮，圆柱形。叶片长圆形或长圆状椭圆形，长25~60cm，先端短尾尖，基部楔形或圆楔形，两侧稍不对称；叶柄长10~30cm，基部具鞘。伞形花序顶生具5~7花；总苞4片，外轮2片卵状披针形，紫色，内轮2片宽卵形，淡绿色；小苞片线形，长10~15cm；花被裂片6，红色或紫褐色；雄蕊6；柱头3浅裂。浆果肉质，具6棱，紫褐色，长约3cm，顶端具宿存花被裂片。花果期4~11月。

产东安、江华、新宁、宜章。生山地溪边、林下阴湿处、田边。分布于广东、广西、海南、贵州、云南、西藏。孟加拉国、印度、老挝、马来西亚、缅甸、斯里兰卡、泰国、越南亦产。全株有毒。

Perennial herbs. Rhizomes stout, terete. Leaf blade oblong or oblong-elliptic, 25-60cm long, apex short caudate, base cuneate or rounded-cuneate, slightly asymmetry; petioles 10-30cm long, base sheathing. Umbels terminal, with 5-7 flowers; involucral bracts 4, outer 2 ovate-lanceolate, purple, inner 2 broadly ovate, pale green; bracteoles linear, 10-15cm long; perianth lobes 6, red or purplish brown; stamens 6; stigma 3-lobed. Berry fleshy, 6-ridged, purplish brown, ca.3cm long, apex with persistent perianth lobes. Fl. and fr. Apr-Nov.

Distributed in Hunan(Dongan, Jianghua, Xinning, Yizhang), Guangdong, Guangxi, Hainan, Guizhou, Yunnan and Tibet, also in Bangladesh, Cambodia, India, Laos, Malaysia, Myanmar, Sri Lanka, Thailand, Vietnam. Grows in streamsides, moist places, cropland edges in mountain land. The whole plant is toxic.

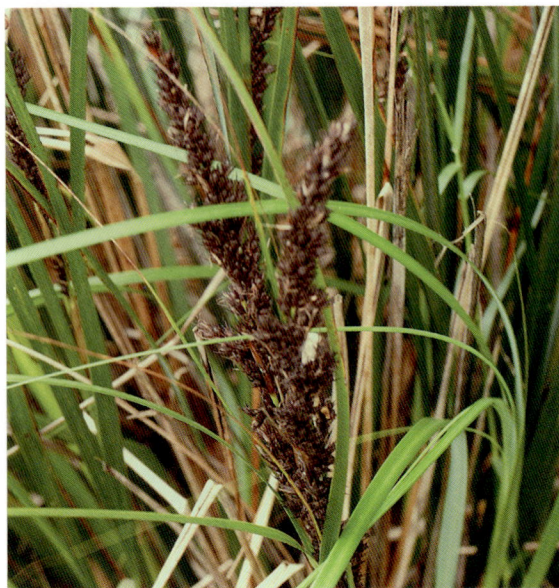

黑莎草 *Gahnia tristis* Nees

暗褐色，卵状披针形，基部6鳞片内无花，最上部2鳞片较小，顶端微凹。小坚果倒卵状长圆形，三棱形，长约4mm，熟时黑色。花果期3～12月。

产宜章。生湿地草丛、林下、路旁。分布于福建、海南、广东、广西、江西。印度、印度尼西亚、日本、马来西亚、泰国、越南亦产。果可榨油。

Perennial herbs, Creeping rhizomes rigid. Culms robust, stiff, 50-150cm tall, terete, hollow, noded. Leaves basal and cauline, leaf sheaths reddish-brown, 10-20cm long; leaf blade strip, subleathery, 40-60cm long, apex acuminate, margin and abaxially spiny serrulate. Inflorescence a contracted panicle, 14-35cm long, with 7-15 oblong spikes; spikelets fusiform, scales 8, initially yellowish brown but maturing to dark brown, ovate-lanceolate, basal 6 scales without flowers, apical 2 scales small, apex retuse. Nutlet obovate-oblong, prismatoidal, ca.4mm long, black when mature. Fl. and fr. Mar-Dec.

Distributed in Hunan(Yizhang), Fujian, Hainan, Guangdong, Guangxi and Jiangxi, also in India, Indonesia, Japan, Malaysia, Thailand, Vietnam. Grows in wet grass, forests, roadsides. Fruit can be extracted oil.

多年生草本。匍匐根状茎坚硬。秆粗壮，坚实，高50～150cm，圆柱状，空心，有节。叶基生和秆生，叶鞘红棕色，穗状，长10～20cm；叶片条形，近革质，长40～60cm，先端渐尖，边缘及背面具刺状细齿。圆锥花序紧缩，长14～35cm，由7～15个长圆形穗状花序组成；小穗纺锤形，鳞片8，初时黄棕色，熟时

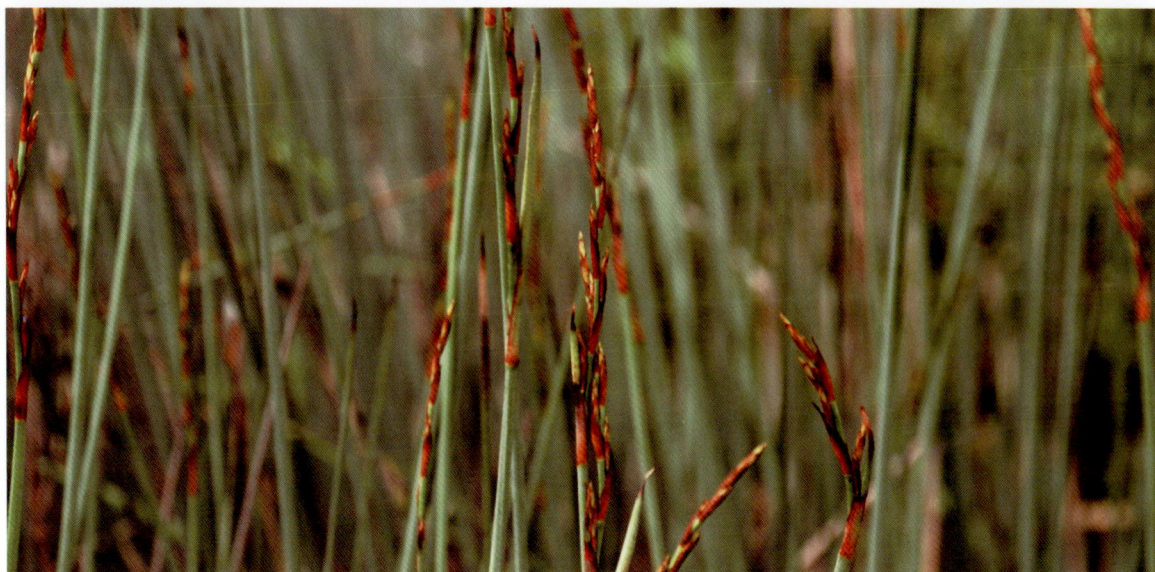

多年生草本，具匍匐根状茎。秆丛生，直立，坚挺，高45～90cm，圆柱状或具不明显棱，基部被枯萎的紫黑色叶鞘。叶基生，圆柱状，较秆稍短，径2～3mm，坚挺，平滑。花序圆锥状，长3～10cm；小穗密集丛生，纺锤状长圆形，长6～8mm；鳞片5，卵形或卵状披针形，顶端钝，具短尖，最下部2鳞片内无毛，最上面1鳞片不发达。小坚果椭圆形，长3.5～4mm，褐黄色，平滑，有光泽。花果期5～12月。

产宜章。生山地疏林下、溪边、路边。分布于福建、广东、海南。印度尼西亚、马来西亚、巴布亚新几内亚、越南亦产。

Perennial herbs, with creeping rhizomes. Culms tufted, erect, stiff, 45-90cm tall, terete or obscurely angular, withered purple sheath at base. Leaves basal, terete, slightly shorter than culms, 2-3mm in diam., stiff, smooth. Inflorescences paniculate, 3-10cm long; spikelets densely clustered, fusiform-oblong, 6-8mm long; scales 5, ovate or ovate-lanceolate, apex obtuse and mucronate, lower 2 scales inside glabrous, top 1 scale not developed. Nutlet oblong, 3.5-4mm long, brownish yellow, smooth, shiny. Fl. and fr. May-Dec.

Distributed in Hunan(Yizhang), Fujian, Guangdong and Hainan, also in Indonesia, Malaysia, Papua New Guinea, Vietnam. Grows in sparse forests, streamsides, roadsides in mountain land.

中文名索引

拉丁学名索引